PHARMACEUTICAL APPLICATIONS OF POLYMERS

a selected bibliography

edited by Richard Juniper

© 1981   ISBN 0–902348–23–X
Rubber and Plastics Research Association of Great Britain,
Shawbury, Shrewsbury, Shropshire, SY4 4NR, England
Telephone: Shawbury (0939)250383 Telex: 35134

CONTENTS

RAPRA

RAPRA is the international technical and information centre for rubbers and plastics. It has continuously developed its functions, skills and resources to meet the needs of the polymer industry for more than 60 years. RAPRA is not a medical organisation but as can be seen from this book, pharmaceutical applications of polymers are an expanding area within RAPRA's traditional fields of interest.

The compiler and editor is a pharmacologist employed at RAPRA as an information specialist. He is also editor of RAPRA's monthly current awareness abstracts bulletin Biomedical Applications of Polymers.

FOREWORD

The view has been expressed that the scope and, indeed, the need for the development of new drugs on the scale that it has proceeded over the past decades is limited. While one may disagree with this view, it is true that in some disease states the stage has been reached where the requirement is not for new drug entities but for optimising the activity of existing drugs by increasing their specificity and decreasing side effects, for example, by controlling the rate of drug release. Pharmaceutical formulation has a role to play in the pursuit of this goal. In recent years the utility of polymers in the fabrication of controlled drug release dosage forms — or drug delivery systems — has been appreciated. Such delivery systems ideally control the rate and extent of drug release from the site of administration of the drug and some systems will direct the drug into the target organ. Many are specifically designed to give prolonged release of therapeutic agent for chronic treatment. Administration of the majority of therapeutic substances will continue in the future to be by tablet, capsule and other conventional dose forms. The advent of potent drugs, active in minute dosages, however, highlighted the need for new initiatives in dosage form design.

Polymers have a major role to play in the fabrication of delivery systems. The variety of polymer structure and form is great; they have the advantage over small molecules in that their molecular weight determines their physical form and function, and there is the possibility of synthesising polymers for specific purpose, or in devising new uses for existing materials.

Such is the diversity of the polymeric and copolymeric entities and such is the variety of the potential uses of polymeric materials that it is rarely possible for one individual to be familiar with the whole range of literature, especially as it is scattered in chemical, polymer, pharmaceutical and medical journals. Hence the need for a work such as this by Richard Juniper.

The publication of this comprehensive compilation of references on polymers used in pharmacy and medicine is opportune for it comes at a time when interest in the topic is burgeoning. In these days of computerised literature searches and abstract journals, it might be thought that a collection such as this is redundant, but as each paper has been considered and arranged for inclusion by Mr. Juniper, the value of the information the collection contains is increased beyond that of a simple listing. The index of polymers and copolymers, the author index and the active ingredient index enhances its usefulness as a reference source. The book will be of value not only to researchers and others who are beginning work in this area, but also to established workers in the field.

It is to be hoped that perusal of this collection will stimulate research workers to develop some of the many themes which appear in the many papers listed here. Mr. Juniper is to be congratulated for initiating the work of collation and seeing it to fruition. I hope that now the ground work has been laid we can look forward to a regular updating of the collection.

A.T. Florence,
Glasgow, 1981

INTRODUCTION

Pharmaceutical applications of synthetic polymers, in particular the delivery of pharmacologically active agents, found a new impetus in the 1960s with reports of sustained drug release from silicone rubber. A number of techniques have since been analysed and developed using new biocompatible synthetic polymers to incorporate drugs or enzymes for achieving their controlled and generally sustained release.

Drug plasma levels of any patient at a particular time depends on compliance with a prescribed dosage routine. For any drug there is an upper plasma level above which toxic effects are seen, and a lower plasma level below which the drug is ineffectual. Controlled release dosage forms aim at achieving a desirable level of drug efficacy providing a consistent level of medication without undesirable toxic effects.

In the context of this book a controlled release drug delivery system is a combination of a pharmacologically active agent and a synthetic polymer designed in such a manner as to release the drug in a biological environment at a constant or slowly declining rate. Drug release occurs either by erosion of the polymer by physical means such as dissolution, by chemical processes such as hydrolysis of the polymer backbone or crosslinks, or by diffusion. If dimensions are constant for a membrane-bound drug delivery system, and providing the internal drug remains saturated and the external medium is unsaturated, a concentration gradient is set up and a constant rate of diffusion occurs from the device. The rate of drug release can vary with the shape of the system and the composition of the polymer. Devices based on this principle are commercially available and include an intrauterine contraceptive system which maintains a constant release of progesterone and an ocular insert which controls the release of pilocarpine in the treatment of glaucoma. Monolithic therapeutic systems have been manufactured in which the drug is homogeneously dissolved or dispersed in the polymer so that the rate of drug release decreases rapidly with time. Where the drug is in solution the rate decays exponentially with time whereas in the case of matrix entrapment the rate of drug release decreases with time and is directly proportional to $t^{\frac{1}{2}}$.

An important criterion in the design of drug carrier systems, in which a drug is covalently attached to the polymer backbone, is the molecular weight of the polymer. This factor can determine whether or not the drug will cross the blood-brain barrier, get excreted by the kidneys or accumulate in certain organs such as the liver or spleen.

Most of the early work with controlled release devices involved subcutaneous implants of silicone rubber-encapsulated steroids. The main disadvantage of using implants made of a non-biodegradable material is that they require surgery for removal once the drug is completely released. A further disadvantage is that a fibrous capsule can build up around the implant which may decrease the rate of diffusion of the drug from the implant. The polymer in diffusible devices has to be permeable to the drug and permeability must not be adversely affected by the immediate environment of the system. The use of erodable polymers such as polyglycolic acid and polylactic acid can help to reduce these problems. Biodegradable drug delivery systems can be roughly divided into three types of polymer:

(1) water-soluble polymers made insoluble by hydrolytically unstable crosslinks;
(2) water-insoluble polymers that are made soluble by hydrolysis but maintain their backbone structure, and
(3) water-insoluble polymers that become soluble by backbone cleavage

An erodable polymer need not be permeable to the drug but must degrade at an appropriate rate providing a constant rate of drug release. The degradation products of the polymer must be completely metabolised with no toxic side effects. Physical forms of dosage include capsules, pellets or discs that are surgically implanted; injectable microcapsules; envelopes, films or laminates that are inserted in the eye or oral cavity or implanted subcutaneously.

The attachment of pharmacologically active agents to the backbone of an inactive polymeric chain has been found to produce a pharmacologically active biopolymer where the system is biodegradable. Transport mechanisms for polymer-drug carriers can be made specific for certain cells by the attachment of so called 'homing devices' such as receptor-active components with pH sensitive groups. Carboxylates, quaternary amines or sulphonates can be added to alter the hydrophilic nature of the system in aqueous media whereas alkyl groups alter solubility in lipid regions. Methods of drug attachment to polymer backbones are dependent on the drug target. Catecholamines bound directly to polyacrylic acid have been found to be ineffective in the physiological environment and yet were found to be active when suspended away from the polymer backbone. Isoprenaline was found to have pharmacological activity only when coupled to a polymer chain by pendant azo groups but not when directly attached to the chain.

Hydrogels have been increasingly used as vehicles for the immobilisation, encapsulation and controlled release of substances such as antibiotics, anticancer drugs, antibodies, enzymes and antibacterial agents. The water swollen network structure of hydrogels enable biologically active compounds to be either physically entrapped or covalently bound within the gel to give either temporary or permanent immobilisation. Hyrdogels are specially suitable for the release of drugs such as antibiotics into tissue and spaces with primary and secondary infection as they permit continuous release of drug into the immediate area to obtain optimum concentration. Such systems are now well established as ocular insert devices for treatment of the eye, for the local release of cancer chemotherapeutic agents and for the controlled release of fluoride for dental applications.

Crosslinked polymers of 2-hydroxyethyl methacrylate (poly HEMA, Hydron) are the hydrogels most commonly used in biomedical applications because of their stability to varying conditions of pH, temperature and toxicity. PolyHEMA has been used as a vehicle in an attempt to produce *in vivo* sustained release of narcotic antagonists over a long period. By altering the nature of the polymer a zero order diffusion rate could be obtained and used in the treatment of drug addiction. Polyvinylpyrrolidone (PVP) has been used with silicone rubber and polyacrylamide for the sustained release of prostaglandins for possible use in obstetrics.

Various enzymes can be attached to polymers and still retain their physiological activity. One such system can be formed by polymerising L-asparaginase, acrylamide and N_1N^2-methylene bis acrylamide. This gives a copolymer matrix containing embedded L-asparaginase which retains its enzymatic activity longer than free L-asparaginase and is more effective in the treatment of leukemia and lymphoma. Immobilised enzymes have also been used to investigate the action of an artificial kidney for the treatment of acatalasaemia, as digestive aids, and as a monitor of glucose levels in conjunction with insulin infusion systems.

Some synthetic polymers do themselves exhibit pharmacological activity, perhaps the most well known examples being divinylether-maleic anhydride copolymer (Pyran) as an inducer of interferon production by the body's immune response, and polyvinylpyridine oxide in the treatment of silicosis. Urea-formaldehyde copolymer (Anaflex) has been used as an anti-microbial agent in the treatment of furunculosis, acne, urinary dermatitis and fungal infections in humans. PVP has been used as a blood plasma extender and the aqueous solution of its iodine complex is marketed as an antiseptic. Polyethylene oxide phenolic ethers have been shown to suppress experimental tuberculosis. However, structure-activity relationships with reference to the number of ethylene oxide units and the nature of the phenolic ethers were shown to affect the mechanism of action. Antitubercular activity was found to be greatest with 15 to 20 ethylene oxide units but with 45–75 units there was no chemotherapeutic effect and the infection was enhanced. This would seem to reinforce opinion that any polymer used in biological systems must undergo the most thorough analysis for toxicity and biological activity for they or their metabolites may possess their own pharmacological activity.

In view of the many and varied developments in the pharmaceutical use of synthetic polymers a need has been felt for a number of years for a basic reference source to bring together the fragmented literature in this field. This book aims to provide such a source for research scientists and for companies developing products in this area. No guide of this kind can hope to be exhaustive, but every effort has been made to include all important English language references and to describe them accurately up to the end of December 1980. Patent literature is not included as this area is covered adequately elsewhere by such organisations as the Noyes Data Company and Lexington Data.

ACKNOWLEDGEMENTS

I wish to express my appreciation and thanks to members of the library staff at the Barnes Medical Library, University of Birmingham; the Medical Library, University of Manchester; University College Hospital Medical School Library and School of Pharmacy Library, University of London; and the British Library Lending Division for their help, and to Elaine Cooper, RAPRA Librarian, for her invaluable assistance. I am indebted to Professors A.T. Florence and N.B. Graham of the University of Strathclyde for their advice and encouragement.

PREFACE

References are numbered and arranged in alphabetical order of polymer or copolymer. Where possible, abbreviations and trade names are also given. Where more than one polymer is cited the reference also appears under the other polymer(s). Further division of polymer is made chronologically with the earliest reference first. Within a given year references are listed alphabetically by author.

Each reference includes the full title, first and second authors' surname and initials (where there are more than two authors the surname and initials of the first author is recorded plus et al), journal title, volume number, year and title page number. When not mentioned in the title the pharmacologically active agent and method of polymer application, e.g. microcapsule or implant, etc., are listed under the reference.

Further references might have been included were it not for the fact that polymers used in some experiments were not described. The most common form of wording was 'a controlled-release preparation was used' or 'embedded in an inert plastic matrix' whilst leaving out the name of the polymer or its chemical structure.

An alphabetical index is given listing polymers and copolymers referred to in the bibliography. Subdivisions of the polymer class are arranged under the main polymer heading. The number after each polymer indicates their position in the bibliography. Alphabetical indices are also given for authors and active agents.

GENERAL

1

ION EXCHANGE AND ADSORPTION AGENTS IN MEDICINE
Martin, G.J.
Little, Brown & Co. 1955

2

ORAL PROLONGED ACTION MEDICAMENTS: THEIR PHARMACEUTICAL CON-
TROL AND THERAPEUTIC ASPECTS
Lazarus, J. & Cooper, J.
J. Pharm. Pharmacol., 11, 1959, p. 257

3

PLASTICS IN PHARMACEUTICAL PRACTICE AND RELATED FIELDS
Autian, J.
J. Pharm. Sci., 52, 1963, p. 1

4

PLASTICS IN PHARMACEUTICAL PRACTICE AND RELATED FIELDS. II
Autian, J.
J. Pharm. Sci., 52, 1963, p. 105

5

METABOLIC INHIBITORS
Bernfeld, P.
Academic Press, N.Y., Vol. 2, 1963, p. 437
Hochster, R.M. & Quastel, J.H. eds.

6

DISSOLUTION RATES OF FINELY DIVIDED DRUG POWDERS. I. EFFECT OF A
DISTRIBUTION OF PARTICLE SIZES IN A DIFFUSION-CONTROLLED PROCESS
Higuchi, W.I. & Hiestand, E.N.
J. Pharm. Sci., 52, 1963, p. 67
PHARMACOKINETICS

7

DISSOLUTION RATES OF FINELY DIVIDED DRUG POWDERS. II. MICRONISED
METHYLPREDNISOLONE
Higuchi, W.I. et al.
J. Pharm. Sci., 52, 1963, p. 162

8

MECHANISM OF SUSTAINED ACTION MEDICATION. THEORETICAL ANALYSIS OF RATE OF RELEASE OF SOLID DRUGS DISPERSED IN SOLID MATRICES
Higuchi, T.
J. Pharm. Sci., 52, 1963, p. 1145

9

CLINICAL EVALUATION OF SUSTAINED RELEASE DRUGS
Frederik, W.S. & Cass, L.J.
J. New Drugs, 5, 1965, p. 138

10

CLINICAL PHARMACOLOGY OF PROLONGED RELEASE DRUGS
Silson, J.E.
Drug Cosmetic Ind., 96, 1965, p. 632

11

MATHEMATICAL MODEL OF SUSTAINED-RELEASE PREPARATIONS AND ITS ANALYSIS
Kruger-Thiemer, E. & Eriksen, S.P.
J. Pharm. Sci., 55, 1966, p. 1249

12

THEORETICAL FORMULATION OF SUSTAINED-RELEASE DOSAGE FORMS
Robinson, J.R. & Eriksen, S.P.
J. Pharm. Sci., 55, 1966, p. 1254

13

WATER-INSOLUBLE DERIVATIVES OF ENZYMES, ANTIGENS AND ANTIBODIES
Silman, I.H. & Katchalski, E.
Ann. Rev. Biochemistry, 35, 1966, p. 873

14

CLINICAL INVESTIGATION IN DRUGS DESIGNED FOR PROLONGED RELEASE
Silson, J.E.
J. New Drugs, 6, 1966, p. 208

15

DIFFUSIONAL MODELS USEFUL IN BIOPHARMACEUTICS
Higuchi, W.I.
J. Pharm. Sci., 56, 1967, p. 315

16

DELAYED RELEASE FORMULATION: IN VIVO EVALUATION
Beckett, A.H.
Pharm. J., 201, 1968, p. 425

17

UTILISATION OF POLYELECTROLYTE COMPLEXES IN BIOLOGY AND MEDICINE
Markley, L.T. et al.
J. Biomed. Mat. Res., 2, 1968, p. 145

18

ORAL PROLONGED RELEASE PHARMACEUTICAL PREPARATIONS
Nairn, G.
Can. Pharm. J., 102, 1969, p. 336

19

ORAL PROLONGED RELEASE PHARMACEUTICAL PREPARATIONS
Nairn, G.
Can. Pharm. J., 102, 1969, p. 368

20

GROWTH REGULATING ACTIVITY OF POLYANIONS: A THEORETICAL
DISCUSSION OF THEIR PLACE IN THE INTERCELLULAR ENVIRONMENT AND
THEIR ROLE IN CELL PHYSIOLOGY
Regelson, W.
Adv. Cancer Res., 11, 1968, p. 223

21

ANTIMITOTIC ACTIVITY OF POLYANIONS
Regelson, W.
Adv. Chemother., 3, 1968, p. 303

22

SUSTAINED RELEASE PHARMACEUTICALS
Williams, A.
Noyes Development Corporation, 1969
BASED ON U.S. PATENTS

23

MICROENCAPSULATION
Luzzi, L.A.
J. Pharm. Sci., 59, 1970, p. 1367

24

PHARMACEUTICAL APPLICATIONS OF SOLID DISPERSION SYSTEMS
Chiou, W.L. & Riegelman, S.
J. Pharm. Sci., 60, 1971, p. 1281

25
EFFECT OF THE MICROENVIRONMENT ON THE MODE OF ACTION OF
IMMOBILISED ENZYMES
Katchalski, E. et al.
Adv. Enzymology, 34, 1971, p. 445

26
PROGRAMMED DRUG RELEASE FROM ORAL DOSAGE FORMS
Lehmann, K.
Pharma Int., Engl. Ed., No. 3, 1971, p. 34

27
PERORAL SOLID DOSAGE FORMS WITH PROLONGED ACTION
Ritschel, W.A.
Pharma Int., Engl. Ed., No. 3, 1971, p. 18

28
STUDIES ON A SUSTAINED RELEASE PRINCIPLE BASED ON AN INERT PLASTIC
MATRIX
Sjogren, J.
Acta Pharm. Suec., 8, 1971, p. 153

29
SUSTAINED DELIVERY OF DRUGS FROM POLYMER/DRUG MIXTURES
Yolles, S. et al.
Polym. News, 1, 1971, p. 9

30
MICROENCAPSULATION OF DRUGS
Bakan, J.A. & Sloan, F.D.
Drug Cosmetic Ind., 110, March 1972, p. 34

31
ORGANIC POLYMER BIOCOMPATIBILITY AND TOXICOLOGY
Bischoff, F.
Clinical Chemistry, 18, 1972, p. 869

32
NEW OCULAR INSERT DEVICE FOR CONTINUOUS CONSTANT-RATE DELIVERY
OF MEDICATION TO THE EYE
Dohlman, C.H. et al.
Ann. Ophthalmol., 4, 1972, p. 823
HYDROCORTISONE

33
INTRAUTERINE CONTRACEPTION
Tatum, H.J.
Am. J. Obstet. Gynecol., 112, 1972, p. 1000

34
DISSOLUTION OF ALKYL VINYL ETHER-MALEIC ANHYDRIDE COPOLYMERS AND
ESTER DERIVATIVES
Woodruff, C.W. et al.
J. Pharm. Sci., 61, 1972, p. 1916

35
NEW APPROACH TO THE TECHNIQUES OF DRUG ADMINISTRATION
Zaffaroni, A.
Manuf. Chem. Aerosol News, 43, Feb. 1972, p. 19

36
LONG TERM EFFECT OF PLASTICS
Anon
Food Cosmet. Toxicol., 10, 1972, p. 567

37
NEW FIELD OF PLASTICS TOXICOLOGY — METHODS AND RESULTS
Autian, J.
Critical Rev. Toxicol., 2, 1973, p. 1

38
SUSTAINED RELEASE SYSTEMS FOR FERTILITY CONTROL
Duncan, G.W. & Kalkwarf, D.R.
Human Reproduction, Conception and Contraception,
1973, p. 483
Hafez, E.S.E. & Evans, T.N. eds., Harper & Row,

39
SUSTAINED RELEASE HORMONAL PREPARATIONS. A SURVEY OF THE FIELD
Kincl, F.A. & Rudel, H.W.
Excerpta Med. Int. Congr. Ser., Series 273, 1973, p. 977

40
BIOLOGIC ACTIVITY OF WATER-SOLUBLE POLYMERS
Regelson, W.
Polymer Sci. Technol., 2, 1973, p. 161

41
INTRAUTERINE DEVICES: CLINICAL ASPECTS
Tietze, C.
Human Reproduction, Conception and Contraception, 1973, p. 293
Hafez, E.S.E. & Evans, T.N. eds., Harper & Row

42
IMMOBILISED ENZYMES
Zaborsky, O.
CRC Press, Cleveland, Ohio, 1973

43
MICROCAPSULE DRUG DELIVERY SYSTEMS
Bakan, J.A.
Polym. Sci. Technol., 8, 1974, p. 213
Johnson & Johnson Symposium, N.J., July 1974
REVIEW

44
CONTROLLED RELEASE: MECHANISM AND RATES
Baker, R.W. & Lonsdale, H.K.
Controlled Release of Biologically Active Agents
Adv. Expl. Med. Biol., 47, 1974, p. 15
Tanquary, A.C. & Lacey, R.E. eds.

45
SPECIAL REQUIREMENTS FOR PLASTIC DEVICES
Bischoff, F. & Bryson, G.
Acta Endocrinol., 75, 1974, Suppl. 185
CONTRACEPTIVE DEVICES

46
INFLUENCE OF SHAPE FACTORS ON KINETICS OF DRUG RELEASE FROM
MATRIX TABLETS. I. THEORETICAL
Cobby, J. et al.
J. Pharm. Sci., 63, 1974, p. 725

47
INTRODUCTION TO CONTROLLED RELEASE
Cowsar, D.R.
Controlled Release of Biologically Active Agents
Adv. Expl. Med. Biol., 47, 1974, p. 1
Tanquary, A.C. & Lacey, R.E. eds.

48
DRUG DELIVERY SYSTEMS: DESIGN CRITERIA
Cowsar, D.R.
Polym. Sci. Technol., 8, 1974, p. 237

49
WHAT GOOD ARE MICROCAPSULES?
Fanger, G.O.
Chemtech., July 1974, p. 397

50
PHARMACEUTICAL USES OF POLYMERS: SOME RECENT DEVELOPMENTS
Florence, A.T.
Pharm. J., 213, 1974, p. 36

51
INFLUENCE OF PHYSICO-CHEMICAL PROPERTIES OF DRUG AND SYSTEM ON
RELEASE OF DRUGS FROM INERT MATRICES
Flynn, G.L.
Controlled Release of Biologically Active Agents
Adv. Expl. Med. Biol., 47, 1974, p. 73
Tanquary, A.C. & Lacey, R.E. eds.

52
MASS TRANSPORT PHENOMENA AND MODELS. THEORETICAL CONCEPTS
Flynn, G.L. et al.
J. Pharm. Sci., 63, 1974, p. 479

53
POLYMERIC CHEMOTHERAPEUTIC AGENTS WITH CARCINOLYTIC ACTIVITY
Iliev, I. et al.
Russ. Chem. Rev., 43, 1974, p. 69

54
ADMINISTRATION OF DRUGS VIA IUDs
Laufe, L.E.
Intrauterine Devices: Development, Evaluation and Program Im-
plementation, 1974, p. 243
Wheeler, R.G. et al. eds.
Academic Press

55
THERAPEUTIC SYSTEMS FOR CONTROLLED ADMINISTRATION OF DRUGS: A
NEW APPLICATION OF MEMBRANE SCIENCE
Michaels, A.S.
ACS Coatings & Plastics Preprints, 34, No. 1, 1974, p. 559

56
DIFFUSIONAL SYSTEMS FOR CONTROLLED RELEASE OF DRUGS TO THE EYE
Shell, J.W. & Baker, R.W.
Ann. Ophthalmol, 6, 1974, p. 1037
PILOCARPINE

57
MICROENCAPSULATION: PROCESSES AND APPLICATIONS
Vandegaer, J.E. ed.
Plenum, 1974

58
BIOMEDICAL APPLICATIONS OF POLYMERIC MATERIALS AND THEIR IN-
TERACTIONS WITH BLOOD COMPONENTS: A CRITICAL REVIEW OF CURRENT
DEVELOPMENTS
Bruck, S.D.
Polymer, 16, 1975, p. 409

59
SOLUTION-SOLUBILITY DEPENDENCY OF CONTROLLED RELEASE OF DRUG
FROM POLYMER MATRIX: MATHEMATICAL ANALYSIS
Chien, Y.W. et al.
J. Pharm. Sci., 64, 1975, p. 1643

60
SYNTHETIC BIOLOGICALLY ACTIVE POLYMERS
Donaruma, L.G.
Prog. Polym. Sci., 4, 1975, p. 1

61
THERMODYNAMIC METHOD OF PREDICTING THE TRANSPORT OF STEROIDS IN
POLYMER MATRICES
Michaels, A.S. et al.
Amer. Inst. Chem. Eng. J., 21, 1975, p. 1073
PREGNADIENE METHYLDIONE; PROGESTERONE; NORPROGESTERONE;
TESTOSTERONE; PREGNATRIENE METHYL ETHINYLACETATE; ESTRADIOL;
NORGESTREL; NORETHINDRONE; CORTISOL; PREDNISOLONE; ESTRIOL

62
STRUCTURE AND PROPERTIES OF PHARMACOLOGICALLY ACTIVE POLYMERS
Ringsdorf, H.
J. Polym. Sci. Polym. Symp., No. 51, 1975, p. 135

63
SUSTAINED RELEASE PLASTIC MATRIX TABLETS
Rowe, R.C.
Manuf. Chem. Aerosol News, 46, March 1975, p. 23

64
DEVELOPMENT OF DEPOT CONTRACEPTIVES
Shearman, R.P.
J. Steroid Biochem., 6, 1975, p. 899

65
ENGINEERING DEVELOPMENT OF THERAPEUTIC SYSTEMS. A NEW CLASS OF
DOSAGE FORMS FOR THE CONTROLLED DELIVERY OF DRUGS
Yates, F.E. et al.
Adv. Biomed. Engng., 5, 1975, p. 1

66
ERODIBLE CONTROLLED RELEASE SYSTEMS
Baker, R.W. & Lonsdale, H.K.
ACS Coatings & Plastics Preprints, 36, No. 1, 1976, p. 235
MECHANISMS AND KINETICS

67
CONTROLLED RELEASE OF EFFECTORS FROM POLYMERS AND MICROCAPSULES
Banker, G.S.
Midland Macromol. Inst. Polymeric Delivery Systems, 5th Int. Symp., Aug.
1976, p. 25

68
STATUS OF DELIVERY SYSTEMS FOR DRUGS AND FOOD ADDITIVES AS VIEWED
BY FDA
Casola, A.R.
Midland Macromol. Inst., Polymeric Delivery Systems, 5th Int. Symp., Aug.
1976, p. 293

69
ENZYMES IMMOBILISED BY MICROENCAPSULATION WITHIN SPHERICAL
ULTRATHIN POLYMERIC MEMBRANES
Chang, T.M.S.
J. Macromol. Sci. Chem., A10, 1976, p. 245

70
THERMODYNAMICS OF CONTROLLED RELEASE FROM POLYMERIC DELIVERY
SYSTEMS
Chien, Y.W.
ACS Coatings & Plastics Preprints, 36, 1976, p. 326
KINETICS

71

UNIFIED MATHEMATICAL MODEL FOR DIFFUSION FROM DRUG-POLYMER COM-
POSITE TABLETS
Fu, J.C. et al.
J. Biomed. Mater. Res., 10, 1976, p. 743

72

PHARMACY, PHARMACEUTICALS AND MODERN DRUG DELIVERY
Higuchi, T.
Amer. J. Hosp. Pharm., 33, 1976, p. 795

73

POLYMER GRAFTS IN BIOCHEMISTRY
Hixson, H.F. & Goldberg, E.P. eds.
Dekker, New York, 1976

74

CONTROLLED RELEASE FROM ERODIBLE SLABS, CYLINDERS AND SPHERES
Hopfenberg, H.B.
ACS Coatings & Plastics Preprints, 36, No. 1, 1976, p. 229
c.f. ACS Symp. Series, 33, 1976, p. 26
RELEASE KINETICS

75

THERMODYNAMIC METHOD OF PREDICTING THE TRANSPORT OF STEROIDS IN
POLYMER MATRICES
Michaels, A.S. et al.
ACS Coatings & Plastics Preprints, 36, 1976, p. 228

76

MACROMOLECULES AND PHARMACEUTICALS
Michaels, A.S.
Polymer Preprints, 17, Apr. 1976, p. 236

77

DIFFUSIONAL RELEASE OF A SOLUTE FROM A POLYMER MATRIX
Paul, D.R. & McSpadden, S.K.
J. Memb. Sci., 1, 1976, p. 33

78

POLYMERS IN CONTROLLED RELEASE TECHNOLOGY
Paul, D.R.
ACS Symp. Series, 33, 1976, p. 1
Controlled Release Polymeric Formulations

79
SUSTAINED ACTION DOSAGE FORMS
Robinson, M.
Theory and Practice of Industrial Pharmacy, Lea & Febiger, 1976, p. 439
Lachman, L. et al. eds.

80
STRUCTURAL FACTORS GOVERNING CONTROLLED RELEASE
Rogers, C.E.
ACS Coatings & Plastics Preprints, 36, 1976, p. 225
c.f. ACS Symp. Series, 33, 1976, p. 15

81
IMPORTANCE OF SOLUTIONING PARTITIONING ON THE KINETICS OF DRUG
RELEASE FROM MATRIX SYSTEMS
Roseman, T.J. & Yalkowsky, S.H.
ACS Coatings & Plastics Preprints, 36, No. 1, 1976, p. 321
c.f. ACS Symp. Series, 33, 1976, p. 33

82
IMMOBILISED ENZYME PRINCIPLES
Wingard, L.B. et al. eds.
Applied Biochem. & Bioengng, 1, 1976

83
CONTROLLED RELEASE OF BIOLOGICALLY ACTIVE AGENTS
Yolles, S. et al.
ACS Symposium, 36, Apr. 1976, p. 332

84
CONTROLLED-DELIVERY THERAPY IS HERE
Zaffaroni, A.
Chemtech., December 1976, p. 756

85
POSSIBLE CLASSIFICATION OF CONTROLLED DRUG RELEASE
Bagnall, R.D.
Biomat. Med. Dev. Artif. Org., 5, 1977, p. 355

86
POLYMERIC DRUGS
Batz, H-G.
Adv. Polym. Sci., No. 23, 1977, p. 25

87
ENCAPSULATION OF ENZYMES, CELL CONTENTS, CELLS, VACCINES, ANTIGENS,
ANTISERUM, COFACTORS, HORMONES AND PROTEINS
Chang, T.M.S.
Biomedical Applications of Immobilised Enzymes and Proteins, 1, 1977, p.
69
Chang, T.M.S. ed., Plenum Press, N.Y.

88
THERAPEUTIC SYSTEMS: A NOVEL APPROACH TO CONTROLLED DELIVERY OF
PHARMACEUTICAL PRODUCTS
Damani, N. & Theeuwes, F.
Ind. J. Pharmacy, 39, 1977, p. 1

89
SLOW RELEASE PREPARATIONS
Dizick, F.
Aust. J. Hosp. Pharm., 7, 1977, p. 11

90
POLYMERIC AFFINITY DRUGS FOR CARDIOVASCULAR, CANCER AND
UROLITHIASIS THERAPY
Goldberg, E.P.
ACS Polymer Preprints, 18, No. 1, 1977, p. 575

91
SYNTHETIC POLYMERS IN CHEMOTHERAPY: GENERAL PROBLEMS
Kalal, J. et al.
ACS Polymer Preprints, 18, No. 1, 1977, p. 553

92
IMMOBILISED ENZYMES
Manecke, G.
ACS Polymer Preprints, 18, No. 1, 1977, p. 535

93
MARKETS FOR THERAPEUTIC SYSTEMS
O'Neill, W.P. & Wolman, A.J.
Drug & Cosmet. Ind., 120, 1977, p. 28

94
CONTROLLED CHEMOTHERAPY THROUGH MACROMOLECULES
Zaffaroni, A.
ACS Polymer Preprints, 18, No. 1, 1977, p. 526

95
POLYMERIC DRUGS IN THE CHEMOTHERAPY OF MICROBIAL INFECTIONS
Samour, C.M.
ACS Polymer Preprints, 18, No. 1, 1977, p. 559

96
CONTROLLED-RELEASE MULTIPLE-UNITS AND SINGLE-UNIT DOSES
Bechgaard, H. & Nielsen, G.H.
Drug. Dev. Ind. Pharm., 4, 1978, p. 53

97
THERAPEUTIC SYSTEMS AND CONTROLLED DRUG DELIVERY
Chandrasekaran, S.K. et al.
J. Memb. Sci., 3, 1978, p. 271

98
MATHEMATICAL MODELS FOR THE RELEASE OF DRUGS FROM MATRIX TABLETS
Cobby, J.
J. Biomed. Mater. Res., 12, 1978, p. 627

99
STATISTICAL MODELS FOR CONTROLLED RELEASE MICROCAPSULES: RATIONALE
AND THEORY
Dappert, T. & Thies, C.
J. Memb. Sci., 4, 1978, p. 99

100
DYNAMICS OF A MEMBRANE-MODERATED CONTROLLED RELEASE
Fan, L.T. et al.
Math. Biosci., 39, 1978, p. 213

101
ORAL CONTROLLED RELEASE PREPARATIONS
Garcia, C.R. et al.
Pharm. Acta Helv., 53, 1978, p. 99

102
SURVEY OF CHEMOTHERAPEAUTIC POLYMERS
Gebelein, C.G.
Polym. News, 4, 1978, p. 163

103
POLYMERIC BIOMEDICAL MATERIALS
Gibbons, D.F.
Brit. Polym. J., 10, 1978, p. 232

104
POLYMERIC INSERTS AND IMPLANTS FOR THE CONTROLLED RELEASE OF DRUGS
Graham, N.B.
Brit. Polym. J., 10, 1978, p. 260

105
MICROCAPSULES: THEIR PREPARATION AND PROPERTIES
Kondo, T.
Surface and Colloid Science, 10, 1978, p. 1
Matijevic, E., ed. Plenum Press, N.Y.

106
LABELLED CELLS
Rembaum, A. et al.
Chemtech., March 1978, p. 182

107
DESIGN OF POLYMERIC IMMUNOMICROSPHERES FOR CELL LABELLING AND CELL SEPARATION
Rembaum, A. & Margel, S.
Brit. Polym. J., 10, 1978, p. 275

108
SUSTAINED AND CONTROLLED RELEASE DRUG DELIVERY SYSTEMS
Robinson, J.R. ed. Dekker 1978, containing.

OVERVIEW OF PROLONGED ACTION DRUG DOSAGE FORMS
Ballard, B.E.; p. 1—69, 295 refs.

DRUG PROPERTIES INFLUENCING THE DESIGN OF SUSTAINED OR CONTROLLED DRUG DELIVERY SYSTEMS
Lee, V.H-L. & Robinson, J.R.; p. 71—121, 232 refs.

METHODS TO ACHIEVE SUSTAINED DRUG DELIVERY: THE PHYSICAL APPROACH: IMPLANTS
Chien, Y.W.; p. 211—349, 185 refs.

METHODS TO ACHIEVE SUSTAINED DRUG DELIVERY: THE CHEMICAL AP-PROACH
Sinkula, A.A.; p. 411—555, 310 refs.

METHODS TO ACHIEVE CONTROLLED DRUG DELIVERY: BIOMEDICAL ENGINEERING APPROACH
Chandrasekaran, S.K. et al.; p. 557—93, 38 refs.

PHARMACOKINETIC CONSIDERATIONS IN THE DESIGN OF CONTROLLED AND
SUSTAINED RELEASE DRUG DELIVERY SYSTEMS
Kwan, K.C.; p. 595—629, 90 refs.

MULTIPLE DOSING OF SUSTAINED RELEASE SYSTEMS
Welling, P.G. & Dobrinksa, M.R.; p. 631—716, 20 refs.

109
POLYMERIC DRUGS
Samour, C.M.
Chemtech., Aug. 1978, p. 494

110
ARTIFICIAL CELLS AS DRUG CARRIERS IN BIOLOGY AND MEDICINE
Chang, T.M.S.
Drug Carriers in Biology and Medicine 1979
Gregoriadis, G. ed.

111
CALCULATIONS OF DRUG RELEASE RATES FROM CONTROLLED RELEASE
DEVICES. THE SLAB
Hadgraft, J.
Int. J. Pharm., 2, 1979, p. 177
PHARMACODYNAMICS

112
MICROCAPSULE PROCESSING AND TECHNOLOGY
Kondo, A.
Marcel Dekker Inc., 1979

113
GEL ENTRAPPED MULTISTEP SYSTEMS
O'Driscoll, K.F. & Mercer, D.G.
Contemporary Topics in Polymer Science, 3, 1979, p. 319
Shen, M. ed. Plenum
REVIEW; IMMOBILISED ENZYMES

114
BIOLOGICAL ACTIVITY OF POLYANIONS: PAST HISTORY AND NEW PROSPECTS
Regelson, W.
J. Polym. Sci. Polym. Symp., No. 66, 1979, p. 483

115
BINDING OF ANTICANCER DRUGS TO CARRIER MOLECULES
Zaharko, D.S. et al.
Methods in Cancer Res., 16, 1979, p. 347

116
RATIONALES IN THE DESIGN OF RECTAL AND VAGINAL DELIVERY FORMS OF DRUGS
DeBlaey, C. & Polderman, J.
Drug Design IX, 1980, p. 237

117
POLYMERIC ANTIMICROBIAL DRUGS
Donaruma, L.G. et al.
Biomedical Polymers: Polymeric Materials and Pharmaceuticals for Biomedical Use, 1980, p. 401
Goldberg, E.P. & Nakajima, A. eds., Academic Press

118
SYNTHETIC POLYMERS OF PHARMACOLOGICAL AND BIOMEDICAL INTEREST
Ferruti, P. et al.
La Chimica E L'Industria, 62, 1980, p. 109

119
POLYMERIC DERIVATIVES OF 5-FLUOROURACIL AND 6-METHYLTHIOPURINE. CHEMOTHERAPEUTIC POLYMERS IX
Gebelein, C.G.
ACS Coatings & Plastics Preprints, 42, 1980, p. 422

120
CONTROLLED RELEASE OF BIOLOGICALLY ACTIVE COMPOUNDS FROM BIOERODIBLE POLYMERS
Heller, J.
Biomaterials, 1, 1980, p. 51

121
CONTROLLED RELEASE OF MACROMOLECULES FROM POLYMERS
Langer, R. & Folkman, J.
Biomedical Polymers: Polymeric Materials and Pharmaceuticals for Biomedical Use, 1980, p. 113
Goldberg, E.P. & Nakajima, A. eds. Academic Press

122
DEPOT PREPARATIONS
Lippold, B.C.
Pharmacy Int., 1, 1980, p. 60

123
HYDROGELS IN BIOMEDICAL APPLICATIONS
Pedley, D.G. et al.
Brit. Polym. J., 12, 1980, p. 99
PATENT REVIEW

124
TRANSPORT PHENOMENA IN POLYMERS. POLYMERS FOR SUSTAINED RELEASE
OF MACROMOLECULES
Peppas, N.A. & Korsmeyer, R.
Polym. News, 6, 1980, p. 149

125
POPULAR MATRICES FOR ENZYME AND OTHER IMMOBILISATIONS
White, C.A. & Kennedy, J.F.
Enzyme Microb. Technol., 2, 1980, p. 82

126
BIODEGRADABLE DRUG DELIVERY SYSTEM: A REVIEW
Wood, D.A.
Int. J. Pharm., 7, 1980, p. 1

127
DELIVERING DRUGS
Zaffaroni, A.
Chemtech., Feb. 1980, p. 82

128
THERAPEUTIC IMPLICATIONS OF CONTROLLED DRUG DELIVERY
Zaffaroni, A.
ACS Coatings & Plastics Preprints, 42, 1980, p. 441

ACRYLAMIDE POLYMERS

Acrylamide Polymer

129
ANTIGENS AND ENZYMES MADE INSOLUBLE BY ENTRAPPING THEM INTO
LATTICES OF SYNTHETIC POLYMERS
Bernfeld, P. & Wan, J.
Science, 142, 1963, p. 678
TRYPSIN; PAPAIN; RIBONUCLEASE; CHYMOTRYPSIN; AMYLASE; ALDOLASE

130
EVALUATION OF POLYMERIC MATERIALS. IV. GRANULATING AGENTS FOR
COMPRESSED TABLETS
Willis, C.R. et al.
J. Pharm. Sci., 54, 1965, p. 366

131
PREPARATION AND CHARACTERISATION OF LYOPHILISED POLYACRYLAMIDE
ENZYME GELS FOR CHEMICAL ANALYSIS
Hicks, G.P. & Updike, S.J.
Analyt. Chem., 38, 1966, p. 726
GLUCOSE OXIDASE; LACTIC DEHYDROGENASE;

132
MATRIX-BOUND ENZYMES. I. USE OF DIFFERENT ACRYLIC COPOLYMERS AS
MATRICES
Mosbach, K.
Acta Chem. Scand., 24, 1970, p. 2084
CITRATE SYNTHETASE; TRYPSIN

133
MATRIX-BOUND ENZYMES. II. STUDIES ON A MATRIX-BOUND TWO-ENZYME
SYSTEM
Mosbach, K. & Mattiasson, B.
Acta Chem. Scand., 24, 1970, p. 2093
HEXOKINASE; GLUCOSE-6-PHOSPHATE DEHYDROGENASE

134
MODEL REACTIONS FOR SYNTHESIS OF PHARMACOLOGICALLY ACTIVE
POLYMERS BY WAY OF MONOMERIC AND POLYMERIC REACTIVE ESTERS
Batz, H-G. et al.
Angew. Chem. Int. Edn., 11, 1972, p. 1103

135
CONTROL OF DIABETES WITH POLYACRYLAMIDE IMPLANTS CONTAINING IN-
SULIN
Davis, B.K.
Experientia, 28, 1972, p. 348

136
CONTROL OF HAMSTER FERTILITY WITH PROSTAGLANDIN F2 ALPHA IMPLANTS
Davis, B.K. & Chang, M.C.
Acta Endocrinol, 70, 1972, p. 97

137
REPRODUCTIVE PERFORMANCE OF HAMSTERS WITH POLYACRYLAMIDE IM-
PLANTS CONTAINING ETHINYLOESTRADIOL
Davis, B.K. et al.
Acta Endocrinol, 70, 1972, p. 385

138
USE OF BEAD POLYMERISATION OF ACRYLIC MONOMERS FOR IM-
MOBILISATION OF ENZYMES
Nilsson, H. et al.
Biochim. Biophys. Acta, 268, 1972, p. 253
TRYPSIN

139
CARRIER-BOUND BIOLOGICALLY ACTIVE SUBSTANCES AND THEIR
APPLICATIONS
Orth, H.D. & Brummer, W.
Angew. Chem. Int. Edn., 11, 1972, p. 249
BETA-GALACTOSIDASE; POLYMER SUPPORT

140
GENETIC ENGINEERING, ENZYME IMMOBILISATION AND TRANSPLANTATION
Updike, S.J.
J. Pharm. Educ., 36, 1972, p. 718
CERAMIDE TRIHEXOSILASE; GEL

141
CONTROL OF DIABETES WITH POLYACRYLAMIDE IMPLANTS CONTAINING IN-
SULIN
Davies, B.K.
Experientia, 28, 1972, p. 348

142
ASPECTS OF THREE TYPES OF HYDROGELS FOR BIOMEDICAL APPLICATIONS
Bruck, S.D.
J. Biomed. Mat. Res., 7, 1973, p. 353

143
SOME PROPERTIES OF ALPHA-CHYMOTRYPSIN AND BETA-GALACTOSIDASE SUP-
PORTED IN POLYACRYLAMIDE GELS
Bunting, P.S. & Laidler, K.J.
Can. J. Biochem., 51, 1973, p. 1598

144
EFFECTS OF A SERIES OF NEW SYNTHETIC HIGH POLYMERS ON CANCER
METASTASES
Ferruti, P. et al.
J. Med. Chem., 16, 1973, p. 496

145
IMMOBILISED THREE-ENZYME SYSTEM: A MODEL FOR MICROENVIRONMENTAL
COMPARTMENTATION IN MITOCHONDRIA
Srere, P.A. et al.
Proc. Nat. Acad. Sci., USA, 70, 1973, p. 2534
MALEATE DEHYDROGENASE; CITRATE SYNTHETASE; LACTATE
DEHYDROGENASE; GEL

146
PROSTAGLANDIN DELIVERY BY CERVICAL DILATOR
Balin, H. et al.
J. Reprod. Med., 13, 1974, p. 208
PROSTAGLANDIN E2; PROSTAGLANDIN F2-ALPHA

147
PHARMACOLOGICALLY ACTIVE POLYMERS. 6. SULPHADIAZINE AND BORON
DERIVATIVES AS POTENTIAL CARRIERS FOR POLYMERS INTO CANCER TISSUE
Bartulin, J. et al.
Makromol. Chem., 175, 1974, p. 1007

148
DIFFUSION IN POLYMER GEL IMPLANTS
Davis, B.K.
Proc. Nat. Acad. Sci., USA, 71, 1974, p. 3120
BOVINE SERUM ALBUMIN; IMMUNOGLOBULIN; INSULIN; LUTENISING HOR-
MONE; PROSTAGLANDIN F2-ALPHA; GEL

149
TUMOUR GROWTH AND NEOVASCULARISATION: AN EXPERIMENTAL MODEL
USING THE RABBIT CORNEA
Gimbrone, M.A. et al.
J. Nat. Cancer Inst., 52, 1974, p. 413
GEL CORNEAL IMPLANT

150
SOLUBLE OPHTHALMIC DRUG INSERTS
Maichuk, Y.F.
Lancet, 1, 18 Jan. 1975, p. 173
NEOMYCIN; KANAMYCIN; SULPHAMETHOXYPYRIDAZINE; IDOXURIDINE;
FLORENAL; ATROPINE; PILOCARPINE; DEXAMETHASONE

151
SUSTAINED RELEASE OF MACROMOLECULES FROM POLYMERS
Langer, R.; Folkman, J.
Midland Macromol. Inst., Polymeric Delivery Systems, 5th Int. Symp., Aug.
1976, p. 175
SOYBEAN TRYPSIN INHIBITOR; LYSOZYME; ALKALINE PHOSPHATASE;
CATALASE; INSULIN; HEPARIN

152
POLYMERS FOR THE SUSTAINED RELEASE OF PROTEINS AND OTHER
MACROMOLECULES
Langer, R. & Folkman, J.
Nature, 263, Oct. 28, 1976, p. 797

153
IMPROVED STABILITY OF PROTEINS IMMOBILISED IN MICROPARTICLES
PREPARED BY A MODIFIED EMULSION POLYMERISATION TECHNIQUE
Ekman, B. & Sjoholm, I.
J. Pharm. Sci., 67, 1978, p. 693
HUMAN SERUM ALBUMIN; CARBONIC ANHYDRASE; GEL

154
POLYMERS AS INTERFERON INDUCERS
Levy, H.B.
Polymeric Drugs, 1978, p. 305; Academic Press
Donaruma, L.G. & Vogl, O. eds.

155
SYNTHESIS AND SOME PROPERTIES OF ANTITHROMBOGENIC POLYMERS
Plate, N.A.
Polymeric Drugs, 1978, p. 63; Academic Press
Donaruma, L.G. & Vogl, O. eds.

156
FACTORS INFLUENCING PLATELET CONSUMPTION BY POLYACRYLAMIDE
HYDROGELS
Hanson, S.R. et al.
Ann. Biomed. Engng., 7, 1979, p. 357

157
POLYMERIC DRUG BY DIRECT COPOLYMERISATION, POLYMER OF BETA-
ADRENERGIC ANTAGONIST AND ITS BINDING TO RECEPTOR AND ANTIBODY
Pitha, J. et al.
ACS Polymer Preprints, 20, No. 2, 1979, p. 652
ALPRENOLOL

158
SYNTHESIS AND REACTIONS OF HYDROPHILIC FUNCTIONAL MICROSPHERES
FOR IMMUNOLOGICAL STUDIES
Rembaum, A. et al.
J. Macromol. Sci. Chem., A13, 1979, p. 603

159
USE OF GRAFT COPOLYMERS AS ENZYME SUPPORTS. I. THE IMMOBILISATION
OF BETA-GALACTOSIDASE AND PAPAIN ONTO NYLON-POLYACRYLAMIDE
GRAFT COPOLYMERS
Beddows, C.G. et al.
Polym. Bull., 1, 1979, p. 749

160
USE OF GRAFT COPOLYMERS AS ENZYME SUPPORTS. II. THE IMMOBILISATION
OF BETA-GALACTOSIDASE AND GLUCOSE OXIDASE ON CELLULOSE-
POLYACRYLAMIDE GRAFT COPOLYMERS
Abdel-Hay, F.I. et al.
Polym. Bull., 1, 1979, p. 755

Diethyl Acrylamide Polymer

161
PREPARATION OF IMMOBILISED ENZYMES FROM ACRYLIC MONOMERS UNDER
GAMMA RAY IRRADIATION
Maeda, H. et al.
Biotech. Bioeng., 17, 1975, p. 119
GLUCOAMYLASE; INVERTASE; BETA-GALACTOSIDASE

Ethyl Acrylamide Polymer

162
NEW TYPES OF SYNTHETIC INFUSION SOLUTIONS. I. INVESTIGATION OF THE
EFFECT OF SOLUTIONS OF SOME HYDROPHILIC POLYMERS ON BLOOD
Kopecek, J. et al.
J. Biomed. Mater. Res., 7, 1973, p. 179

Pyrimidinyl Sulphamoylphenyl Acrylamide Polymer

163
PHARMACOLOGICALLY ACTIVE POLYMERS: SULPHADIAZINE AND BORON
DERIVATIVES AS POTENTIAL CARRIERS FOR POLYMERS INTO CANCER TISSUE
Bartulin, J. et al.
Makromol. Chem., 175, 1974, p. 1007

ACRYLATE POLYMERS AND COPOLYMERS

Acrylate Polymer

164
EVALUATION OF POLYMERIC MATERIALS. I. SCREENING OF SELECTED POLYMERS AS FILM COATING AGENTS
Munden, B.J. et al.
J. Pharm. Sci., 53, 1964, p. 395

165
CYCLOPHOSPHAMIDE- AND STEROID HORMONE-CONTAINING POLYMERS AS POTENTIAL ANTICANCER COMPOUNDS
Batz, H.G. et al.
Makromol. Chem., 175, 1974, p. 2229
TESTOSTERONE

166
SUSTAINED RELEASE OF SOME PHENOLS FROM POLYMER MATRICES
Ray, A.R. et al.
Symposium on Industrial Polymers & Radiation, Gujarat, Feb. 1979
PHENOLS; SUBSTITUTED PHENOL ANTIMICROBIALS

Ethyl Acrylate Polymer

167
SOLUBLE OPHTHALMIC DRUG INSERTS
Maichuk, Y.F.
Lancet, 1, 18 Jan. 1975, p. 173
NEOMYCIN; KANAMYCIN; SULPHAMETHOXYPYRIDAZINE; IDOXURIDINE; FLORENAL; ATROPINE; PILOCARPINE; DEXAMETHASONE

168
CONTROLLED RELEASE OF DRUGS FROM HYDROGEL MATRICES
Hosaka, S. et al.
J. Appl. Polym. Sci., 23, 1979, p. 2089
ERYTHROMYCIN; ERYTHROMYCIN ESTOLATE

Ethyl Acrylate – Ethylene Copolymer

169
ANTIMICROBIAL POLYMERS
Ackart, W.B. et al.
J. Biomed. Mat. Res., 9, 1975, p. 55

Ethylene Glycol Diacrylate Polymer

170
CERVICAL HYDROGEL DILATOR: A NEW DELIVERY SYSTEM FOR PROSTAGLAN-
DINS
Akkapeddi, M.K. et al.
Adv. Expl. Med. & Biol., 47, 1974, p. 165
Controlled Release of Biologically Active Agents
Tanquary, A.C. & Lacey, R.E. eds.
PROSTAGLANDIN E2; PROSTAGLANDIN F2-ALPHA; MATRIX

171
CONTROLLED RELEASE OF BIOFUNCTIONAL SUBSTANCES BY RADIATION- IN-
DUCED POLYMERISATION. I. RELEASE OF POTASSIUM CHLORIDE BY
POLYMERISATION OF VARIOUS VINYL MONOMERS
Yoshida, M. et al.
Polymer, 19, 1978, p. 1375
CAPSULE

172
MECHANISM OF IMMOBILISATION OF ENZYMES BY RADIATION-INDUCED
POLYMERISATION OF GLASS-FORMING MONOMERS
Yoshida, M. et al.
J. Macromol. Sci. Chem., A14, 1980, p. 541
GLUCOAMYLASE; ENTRAPMENT

Hydroxyethyl Acrylate Polymer

173
PREPARATION OF IMMOBILISED ENZYMES FROM ACRYLIC MONOMERS UNDER
GAMMA RAY IRRADIATION
Maeda, H. et al.
Biotech. Bioeng., 17, 1975, p. 119
GLUCOAMYLASE; INVERTASE; BETA-GALACTOSIDASE

174
INSULIN PERMEABILITY OF HYDROPHILIC POLYACRYLATE MEMBRANES
Sefton, M.V. & Nishimura, E.
J. Pharm. Sci., 69, 1980, p. 208

Hydroxyethyl Acrylate-Methyl Methacrylate Copolymer

175
DRUG ENTRAPMENT FOR CONTROLLED RELEASE IN RADIATION-POLYMERISED
BEADS
Yoshida, M. et al.
J. Pharm. Sci., 68, 1979, p. 628
POTASSIUM CHLORIDE

Methyl Acrylate Polymer

176
EVALUATION OF POLYMERIC MATERIALS. I. SCREENING OF SELECTED
POLYMERS AS FILM COATING AGENTS
Munden, B.J. et al.
J Pharm. Sci., 53, 1964, p. 395

177
CYCLOPHOSPHAMIDE- AND STEROID HORMONE-CONTAINING POLYMERS AS
POTENTIAL ANTICANCER COMPOUNDS
Batz, H.G. et al.
Makermol. Chem., 175, 1974, p. 2229
TESTOSTERONE

178
CONTROLLED RELEASE OF BIOFUNCTIONAL SUBSTANCES BY RADIATION-
INDUCED POLYMERISATION. I. RELEASE OF POTASSIUM CHLORIDE BY
POLYMERISATION OF VARIOUS VINYL MONOMERS
Yoshida, M. et al.
Polymer, 19, 1978, p. 1375
CAPSULE

179
CONTROLLED SLOW RELEASE OF CHEMOTHERAPEUTIC DRUGS FOR CANCER
FROM MATRICES PREPARED BY RADIATION POLYMERISATION AT LOW TEM-
PERATURES
Kaetsu, I.
J. Biomed. Mat. Res., 14, 1980, p. 185
MITOMYCIN; 5-FLUOROURACIL; BLEOMYCIN

180
CONTROLLED RELEASE OF MULTI-COMPONENT CYTOTOXIC AGENTS FROM
RADIATION POLYMERISED COMPOSITES
Kaetsu, I. et al.
Biomaterials, 1, 1980, p. 17
MITOMYCIN; ADRYAMYCIN; TETRAHYDROFURYL FLUOROURACIL

Methyl Acrylate – Methyl Methacrylate Copolymer

181
DRUG RELEASE FROM METHYL ACRYLATE-METHYL METHACRYLATE COPOLYMER
MATRIX. I. KINETICS OF RELEASE
Farhadieh, B. et al.
J. Pharm. Sci., 60, 1971, p. 209
SODIUM PENTOBARBITAL; METHAPYRILENE; EPHEDRINE; DEXTROMETHORPHAN

182
DRUG RELEASE FROM METHYL ACRYLATE-METHYL METHACRYLATE COPOLYMER
MATRIX. II. CONTROL OF RELEASE RATE BY EXPOSURE TO ACETONE VAPOUR
Farhadieh, B. et al.
J. Pharm. Sci., 60, 1971, p. 212
SODIUM PENTOBARBITAL; METHAPYRILENE; EPHEDRINE; DEXTROMETHORPHAN

183
DRUG RELEASE FROM METHYL ACRYLATE-METHYL METHACRYLATE COPOLYMER
MATRIX. III. SIMULTANEOUS RELEASE OF NONINTERACTING DRUG-EXCIPIENT
MIXTURES
Farhadieh, B.
J. Pharm. Sci., 65, 1976, p. 1333
METHAPYRILENE

184
GLC DETERMINATION OF METHAMPHETAMINE HYDROCHLORIDE IN METHYL
ACRYLATE-METHYL METHACRYLATE SUSTAINED RELEASE TABLETS
Sennello, L.T.
J. Pharm. Sci., 60, 1971, p. 595
METHAMPHETAMINE

Trimethylolpropane Triacrylate Polymer

185
CONTROLLED RELEASE OF BIOFUNCTIONAL SUBSTANCES BY RADIATION-
INDUCED POLYMERISATION. I. RELEASE OF POTASSIUM CHLORIDE BY
POLYMERISATION OF VARIOUS VINYL MONOMERS
Yoshida, M. et al.
Polymer, 19, 1978, p. 1375
CAPSULE

ACRYLIC ACID POLYMERS AND COPOLYMERS

Acrylic Acid Polymer

186
SYNTHETIC POLYELECTROLYTES AS TUMOUR INHIBITORS
Regelson, W. et al.
Nature, 186, No. 4727, 1960, p. 778

187
EVALUATION OF POLYMERIC MATERIALS. II. SCREENING OF SELECTED VINYLS
AND ACRYLATES AS PROLONGED-ACTION COATINGS
Nessel, R.J. et al.
J. Pharm. Sci., 53, 1964, p. 790
AMPHETAMINE SULPHATE

188
EVALUATION OF POLYMERIC MATERIALS. IV. GRANULATING AGENTS FOR
COMPRESSED TABLETS
Willis, C.R. et al.
J. Pharm. Sci., 54, 1965, p. 366

189
INTERACTION OF DRUGS WITH POLYMERS. II. THE PHASE SEPARATION OF
POLYACRYLIC ACID BY CATIONIC DRUGS
Tanaka, N. et al.
Chem. Pharm. Bull., 14, 1966, p. 414
ANILINE HCl; METHYLANILINE HCl; DIMETHYLANILINE HCl; EPHEDRINE HCl;
METHYL EPHEDRINE HCl; PYRIDINE HCl; PYRIDOXINE HCl; PIPERIDINE HCl;
THIAMINE HCl; BENZOYLTHIAMINE DISULPHIDE HCl

190
ACRYLIC RESIN COATINGS FOR THE MANUFACTURE OF DEPOT PREPARATIONS
OF DRUGS
Lehmann, K.
Drugs made in Germany, 10, 1967, p. 115

191
DERIVATIVES OF PROTEINS. II. COUPLING OF ALPHA-CHYMOTRYPSIN TO CAR-
BOXYL-CONTAINING POLYMERS BY USE OF N-ETHYL-5-PHENYLISOXAZOLIUM-
3'-SULPHONATE
Patel, R.P. et al.
Biopolymers, 5, 1967, p. 577

192
ACRYLIC RESIN COATINGS FOR DRUGS. RELATIONSHIP BETWEEN THEIR
CHEMICAL STRUCTURE, PROPERTIES AND APPLICATION POSSIBILITIES
Lehmann, K.
Drugs made in Germany, 71, 1968, p. 34

193
EFFECT OF INTERFERON, POLYACRYLIC AND POLYMETHACRYLIC ACIDS ON
TAIL LESIONS IN VACCINIA-INFECTED MICE
De Clercq, E. & De Somer, P.
Appl. Microbiol., 16, 1968, p. 1314

194
PERMEABLE ACRYLIC RESIN COATINGS FOR THE MANUFACTURE OF DEPOT
PREPARATIONS OF DRUGS. 2. COATING OF GRANULES AND PELLETS AND
PREPARATION OF MATRIX TABLETS
Lehmann, K. & Dreher, D.
Drugs made in Germany, 12, 1969, p. 59
TRIFLUOPERAZINEDIHYDROCHLORIDE

195
ANTIVIRAL ACTIVITY OF POLYACRYLIC AND POLYMETHACRYLIC ACIDS. I.
MODE OF ACTION IN VITRO
De Somer, P. et al.
J. Virol., 2, 1968, p. 878

196
ANTIVIRAL ACTIVITY OF POLYACRYLIC AND POLYMETHACRYLIC ACIDS. II.
MODE OF ACTION IN VIVO
De Somer, P. et al.
J. Virol., 2, 1968, p. 886

197
INTERFERON-STIMULATING AND IN VIVO ANTIVIRAL EFFECTS OF VARIOUS
SYNTHETIC ANIONIC POLYMERS
Merigan, T.C. & Finkelstein, M.S.
Virology, 35, 1968, p. 363
CHEMOTHERAPY

198
PROLONGED ANTIVIRAL PROTECTION BY INTERFERON INDUCERS
De Clercq, E. & De Somer, P.
Proc. Soc. Exp. Biol. & Med., 132, 1969, p. 699
CHEMOTHERAPY

199
MOLECULAR-SCALE DRUG ENTRAPMENT AS A PRECISE METHOD OF CON-
TROLLING DRUG RELEASE. I. ENTRAPMENT OF CATIONIC DRUGS BY POLYMERIC
FLOCCULATION
Goodman, H. & Banker, G.S.
J. Pharm. Sci., 59, 1970, p. 1131
METHAPYRILENE

200
INTERFERON STIMULATION BY LOW MOLECULAR WEIGHT POLYACRYLIC ACIDS
Niblack, J.F.
Ann. N.Y. Acad. Sci., 173, 1970, p. 536

201
SYNTHETIC POLYANIONS PROTECT MICE AGAINST INTRACELLULAR BACTERIAL
INFECTION
Remington, J.S. & Merigan, T.C.
Nature, 226, 25 Apr., 1970, p. 361

202
MOLECULAR SCALE DRUG ENTRAPMENT AS A PRECISE METHOD OF CON-
TROLLED DRUG RELEASE. II. FACILITATED DRUG ENTRAPMENT TO POLYMERIC
COLLOIDAL DISPERSIONS
Rhodes, C.T. et al.
J. Pharm. Sci., 59, 1970, p. 1578
CHLORPHENIRAMINE

203
MOLECULAR SCALE DRUG ENTRAPMENT AS A PRECISE METHOD OF CON-
TROLLED DRUG RELEASE. III. IN VITRO AND IN VIVO STUDIES OF DRUG RELEASE
Rhodes, C.T. et al.
J. Pharm. Sci., 59, 1970, p. 1581
PHENYLEPHRINE; PHENYLPROPANOLAMINE

204
INTERACTION OF DRUGS WITH POLYMERS. III. PHASE SEPARATION OF
POLYACIDS BY O-BENZOYLTHIAMINE DISULPHIDE HYDROCHLORIDE AND
GASTROINTESTINAL ABSORPTION OF O-BENZOYLTHIAMINE DISULPHIDE-
POLYACID COMPLEXES
Tanaka, N. et al.
Chem. Pharm. Bull., 18, 1970, p. 1083

205
INTERFERON-INDUCING POLYCARBOXYLATES: MECHANISM OF PROTECTION AGAINST VACCINIA VIRUS INFECTION IN MICE
Billiau, A. et al.
Infect. Immun., 5, 1972, p. 854

206
EVALUATION OF SELECTED PHYSICAL CONSTANTS OF POLYMERIC FILMS AND PROPOSED KINETICS OF DRUG RELEASE
Sciarra, J.J. & Gidwani, R.N.
J. Pharm. Sci., 61, 1972, p. 754

207
ACRYLIC COATINGS IN CONTROLLED RELEASE TABLET MANUFACTURE
Lehmann, K.
Manuf. Chem. Aerosol News, 44, May 1973, p. 36

208
ACRYLIC COATINGS IN CONTROLLED RELEASE TABLET MANUFACTURE. II.
Lehmann, K.
Manuf. Chem. Aerosol News, 44, June 1973, p. 39

209
RIGID SUPPORTS FOR THE IMMOBILISATION OF ENZYMES
Roger, G.P. et al.
J. Macromol. Sci. Chem., A10, 1976, p. 245
TRYPSIN

210
STUDY OF POLYMER DRUG INTERACTED SYSTEMS AS A PHYSICO-CHEMICAL AP-PROACH IN DRUG DESIGN. I. A STUDY OF DIPHENHYDRAMINE ENTRAPMENT COMPOSITIONS AS A NEW CONTRIBUTION TO CONTROLLED RELEASE ORAL DRUGS
Salib, N.N.
1st Int. Conf. Pharm. Technol., 1977, p. 75

211
STUDY OF POLYMER DRUG INTERACTED SYSTEMS AS A PHYSICO-CHEMICAL AP-PROACH IN DRUG DESIGN. II. PREPARATION AND IN VITRO EVALUATION OF DIPHENHYDRAMINE POLYACRYLATE
Salib, N.N.
1st Int. Conf. Pharm. Technol., 1977, p. 88

212
POLYMERS AS INTERFERON INDUCERS
Levy, H.B.
Polymeric Drugs 1978, p. 305; Academic Press
Donaruma, L.G. & Vogl, O. eds.

213
IMMOBILISED ENZYMES
Manecke, G. & Schlunsen, J.
Polymeric Drugs 1978, p. 39; Academic Press
Donaruma, L.G. & Vogl, O. eds.

214
BIOLOGICAL ACTIVITY OF POLYCARBOXYLIC ACID POLYMERS
Ottenbrite, R.M. et al.
Polymeric Drugs 1978, p. 263; Academic Press
Donaruma, L.G. & Vogl, O. eds.

215
CONTROLLED CHEMOTHERAPY THROUGH MACROMOLECULES
Zaffaroni, A. & Bonson, P.
Polymeric Drugs 1978, p. 1; Academic Press
Donaruma, L.G. & Vogl, O. eds.

216
PREPARATION AND PHARMACOLOGICAL ACTIVITY OF BIOCOMPATIBLE
MEDICINE-CARRIER OLIGOMERS
Bauduin, G. et al.
Chimica e Ind., 62, 1980, p. 421
PARACETEMOL

217
USE OF GRAFT COPOLYMERS AS ENZYME SUPPORTS. THE PREPARATION AND
USE OF POLYETHYLENE-CO-ACRYLIC ACID SUPPORTS
Beddows, C.G. et al.
Polym. Bull., 3, 1980, p. 645
BOVINE SERUM ALBUMIN; PHENYL PROPYLAMINE

218
PHARMACEUTICAL STUDY OF THE COPRECIPITATES OF TETRACYCLINE WITH
ACRYLIC RESINS
Ghanem, A. et al.
Pharm. Acta Helv., 55, 1980, p. 61

219
CONTROLLED RELEASE TABLETS. I. USE OF PELLETS COATED WITH A RETAR-
DING ACRYLATE PLASTIC IN TABLETTING
Juslin, M. et al.
Pharm. Ind., 42, 1980, p. 829
PHENAZONE

Acrylic Acid — Ethylene Copolymer

220
ANTIMICROBIAL POLYMERS
Ackart, W.B. et al.
J. Biomed. Mat. Res., 9, 1975, p. 55

Acrylic Acid — Maleic Anhydride Copolymer

221
BIOLOGICAL ACTIVITY OF ANIONIC POLYMERS
Ottenbrite, R.M. et al.
ACS Polymer Preprints, 18, No. 1, 1977, p. 581

222
COMPARATIVE STUDY OF ANTITUMOUR AND TOXICOLOGIC PROPERTIES OF
RELATED POLYANIONS
Ottenbrite, R. et al.
Polymer, 18, 1977, p. 461

223
STRUCTURE AND BIOLOGICAL ACTIVITIES OF SOME POLYANIONIC POLYMERS
Ottenbrite, R.M.
Anionic Polymeric Drugs, 1980, p. 21
Donaruma, L.G. et al. eds.; Wiley-Interscience

ACRYLONITRILE POLYMER (PAN)

224
EFFECT OF SUBSTRATE CONCENTRATION FOR THE ACTIVITIES OF IMMOBILISED
ENZYMES ON POLYACRYLONITRILE CARRIER RESIN
Handa, T. et al.
Reports on Prog. in Polym. Phys. in Japan, 21, 1978

225
POLYMERIC CINCHONA ALKALOIDS
Kobayashi, N. & Iwai, K.
J. Amer. Chem. Soc., 100, 1978, p. 7071

226
USE OF GRAFT COPOLYMERS AS ENZYME SUPPORTS. IV. THE IMMOBILISATION
OF INVERTASE, PEPSIN, ACID AND ALKALINE PHOSPHATASES AND BOVINE
SERUM ALBUMIN TO HYDROLYSED AND REDUCED NYLON-CO-ACRYLONITRILE
GRAFT COPOLYMERS
Abdel-Hay, F.I. et al.
Polym. Bull., 2, 1980, p. 607

227
IMMOBILISATION OF PROTEINS IN MICROSPHERES OF BIODEGRADABLE
POLYACRYLDEXTRAN
Edman, P. et al.
J. Pharm. Sci., 69, 1980, p. 838
CARBONIC ANHYDRASE; BOVINE SERUM ALBUMIN

228
FUNCTIONAL POLYMERS. III. SYNTHESIS AND PROPERTIES OF NEW POLYMERIC
CINCHONA ALKALOIDS
Kobayashi, N. & Iwai, K.
J. Polym. Sci. Polym. Chem., 18, 1980, p. 223
POWDER

ALLYLDIETHYL AMINE POLYMER

229
NEW ANTAGONIST OF HEPARIN: PARTIALLY N-OXIDISED POLYALLYLDIETHYL-
AMINE
Marchisio, M.A. et al.
Eur. J. Pharmacol., 12, 1970, p. 236

AMIDE POLYMERS

Amide Polymer

230
FURTHER STUDIES OF POLYMERS AS CARCINOGENIC AGENTS IN ANIMALS
Oppenheimer, B.S. et al.
Cancer Res., 15, 1955, p. 333

231
SEMIPERMEABLE MICROCAPSULES
Chang, T.M.S.
Science, 146, 1964, p. 524
ERYTHROCYTE HAEMOLYSATE

232
INTERACTION OF WEAK ORGANIC ACIDS WITH INSOLUBLE POLYAMIDES. II.
STUDY OF SORPTION OF SELECTED WEAK ORGANIC ACIDS BY NYLON 610
Kapadia, A.J. et al.
J. Pharm. Sci., 53, 1964, p. 28
BENZOIC ACID; HYDROXYBENZOIC ACID

233
INTERACTION OF WEAK ORGANIC ACIDS WITH INSOLUBLE POLYAMIDES. I.
SORPTION OF SALICYLIC ACID BY NYLON 66
Kapadia, A.J. et al.
J. Pharm. Sci., 53, 1964, p. 720

234
INTERACTION OF SORBIC ACID WITH AN INSOLUBLE POLYAMIDE
Rodell, M.B. et al.
J. Pharm. Sci., 53, 1964, p. 873

235
INTERACTION OF A GROUP OF LOW MOLECULAR WEIGHT ORGANIC ACIDS
WITH INSOLUBLE POLYAMIDES. I. SORPTION AND DIFFUSION OF FORMIC,
ACETIC, PROPIONIC AND BUTYRIC ACIDS INTO NYLON 66
Berg, H.F. et al.
J. Pharm. Sci., 54, 1965, p. 79

236
FURTHER STUDIES ON THE INTERACTION OF SORBIC ACID WITH AN IN-
SOLUBLE POLYAMIDE
Rodell, M.B. et al.
J. Pharm. Sci., 54, 1965, p. 129

237
TISSUE REACTION INDUCED IN GUINEA PIGS BY PARTICULATE POLYMETHYL
METHACRYLATE, POLYTHENE AND NYLON OF THE SAME SIZE RANGE
Stinson, N.E.
Brit. J. Exp. Pathol., 46, 1965, p. 135

238
SEMIPERMEABLE AQUEOUS MICROCAPSULES: PREPARATION AND PROPERTIES
Chang, T.M.S. et al.
Canad. J. Phys. & Pharmacol., 44, 1966, p. 116
ERYTHROCYTE HAEMOLYSATE

239
SEMIPERMEABLE AQUEOUS MICROCAPSULES WITH EMPHASIS ON
EXPERIMENTS IN AN EXTRACORPOREAL SHUNT SYSTEM
Chang, T.M.S.
Trans. Am. Soc. Artif. Int. Org., 12, 1966, p. 13
UREASE

240
INTERACTION OF A GROUP OF WEAK ORGANIC ACIDS AND PHENOLS WITH
A POLYAMIDE
Rodell, M.B. et al.
J. Pharm. Sci., 55, 1966, p. 1429
BENZOIC ACID; SALICYLIC ACID; HYDROXYBENZOIC ACID

241
SEMIPERMEABLE AQUEOUS MICROCAPSULES (ARTIFICIAL CELLS). V.
PERMEABILITY CHARACTERISTICS
Chang, T.M.S. & Poznansky, M.J.
J. Biomed. Mat. Res., 2, 1968, p. 187
UREA; CREATININE; URIC ACID; CREATINE; GLUCOSE; SUCROSE;
ACETYLSALICYLIC ACID

242
MECHANICAL PROPERTIES OF SEMIPERMEABLE MICROCAPSULES
Jay, A.W.L. & Edwards, M.A.
Canad. J. Phys. & Pharmacol., 46, 1968, p. 731
ERYTHROCYTE HAEMOLYSATE

243
SUSTAINED RELEASE HORMONAL PREPARATIONS. I. DIFFUSION OF VARIOUS
STEROIDS THROUGH POLYMER MEMBRANES
Kincl, F.A. et al.
Steroids, 11, 1968, p. 673
NORPROGESTERONE; PROGESTERONE; TESTOSTERONE; NORETHINDRONE;
ESTRADIOL; MESTRANOL; CORTICOSTERONE; CORTISOL

244
MEMBRANE RESISTANCE OF SEMIPERMEABLE MICROCAPSULES
Jay, A.W.L. & Sivertz, K.S.
J. Biomed. Mat. Res., 3, 1969, p. 577

245
PREPARATION AND EVALUATION OF THE PROLONGED RELEASE PROPERTIES
OF NYLON MICROCAPSULES
Luzzi, L.A. et al.
J. Pharm. Sci., 59, 1970, p. 338
SODIUM PENTOBARBITAL

246
EXPERIMENTAL ERRORS RESULTING FROM UPTAKE OF LIPOPHILIC DRUGS BY
SOFT PLASTIC MATERIALS
Minder, R. et al.
Biochem. Pharmacol., 19, 1970, p. 2179
IMIPRAMINE

247
DRUG-PLASTIC INTERACTIONS. II. SORPTION OF p-HYDROXYBENZOIC ACID
ESTERS BY CAPRAN POLYAMIDE AND IN VITRO BIOLOGICAL ACTIVITY
Patel, N.K. & Nagabhushan, N.
J. Pharm. Sci., 59, 1970, p. 264
METHYLPARABEN; PROPYLPARABEN

248
DRUG-PLASTIC INTERACTIONS. II. SORPTION OF p-HYDROXYBENZOIC ACID
ESTERS BY CAPRAN POLYAMIDE AND IN VITRO BIOLOGIC ACTIVITY
Patel, N.K. & Nagabhushan, N.
J. Pharm. Sci., 59, 1970, p. 264
PARABENS

249
STUDIES ON MICROCAPSULES. V. PREPARATION OF POLYAMIDE MICROCAP-
SULES CONTAINING AQUEOUS PROTEIN SOLUTION
Shiba, M. et al.
Chem. Pharm. Bull., 18, 1970, p. 803

250
EVALUATION OF SELECTED PHYSICAL CONSTANTS OF POLYMERIC FILMS AND
PROPOSED KINETICS OF DRUG RELEASE
Sciarra, J.J. & Gidwani, R.N.
J. Pharm. Sci., 61, 1972, p. 754

251
ENZYMATIC PROPERTIES OF MICROCAPSULES CONTAINING ASPARAGINASE
Mori, T.
Biochim. Biophys. Acta, 321, 1973, p. 653

252
IMMOBILISATION OF BETA-GALACTOSIDASE THROUGH ENCAPSULATION IN WATER-INSOLUBLE MICROCAPSULES
Ostergaard, J.C.W. & Martiny, S.C.
Biotechnol. Bioeng., 15, 1973, p. 561

253
MICROCAPSULES. XIV. EFFECTS OF MEMBRANE MATERIALS AND VISCOSITY OF AQUEOUS PHASE ON PERMEABILITY OF POLYAMIDE MICROCAPSULES TOWARD ELECTROLYTES
Takamura, K. et al.
J. Pharm. Sci., 62, 1973, p. 610

254
IN VIVO EFFECTS OF INTRAPERITONEALLY INJECTED L-ASPARAGINASE SOLUTION AND L-ASPARAGINASE IMMOBILISED WITHIN SEMIPERMEABLE NYLON MICROCAPSULES WITH EMPHASIS ON BLOOD L-ASPARAGINASE, 'BODY' L-ASPARAGINASE, AND PLASMA L-ASPARAGINASE LEVELS
Sui Chong, E.D. & Chang, T.M.S.
Enzyme, 18, 1974, p. 218

255
ARTIFICIAL CELLS
Chang, T.M.S.
Chemtech., Feb. 1975, p. 80
CATALASE; ASPARAGINASE

256
ENZYMES IMMOBILISED BY MICROENCAPSULATION WITHIN SPHERICAL ULTRATHIN POLYMERIC MEMBRANES
Chang, T.M.S.
J. Macromol. Sci. Chem., A10, 1976, p. 245
SEBACOYL CHLORIDE

257
EFFECT OF CROSSLINKING AGENTS ON THE RELEASE OF SODIUM PEN-TOBARBITAL FROM NYLON MICROCAPSULES
De Gennaro, M.D. et al.
ACS Coatings & Plastics Preprints, 36, 1976, p. 376
c.f. ACS Symp. Series, 33, 1976, p. 195
SODIUM PENTOBARBITAL

258
IN VITRO RELEASE OF THERAPEUTICALLY ACTIVE INGREDIENTS FROM POLYMER
MATRIXES
Sciarra, J.J. & Patel, S.P.
J. Pharm. Sci., 65, 1976, p. 1519
BENZOCAINE; CYCLOMETHYCAINE; METHAPYRILENE

259
USE OF GRAFT COPOLYMERS AS ENZYME SUPPORTS. I. THE IMMOBILISATION
OF BETA-GALACTOSIDASE AND PAPAIN ONTO NYLON-POLYACRYLAMIDE
GRAFT COPOLYMERS
Beddows, C.G. et al.
Polym. Bull., 1, 1979, p. 749

260
USE OF GRAFT COPOLYMERS AS ENZYME SUPPORTS. IV. THE IMMOBILISATION
OF INVERTASE, PEPSIN, ACID AND ALKALINE PHOSPHATASES AND BOVINE
SERUM ALBUMIN TO HYDROLYSED AND REDUCED NYLON-CO-ACRYLONITRILE
GRAFT COPOLYMERS
Abdel-Hay, F.I. et al.
Polym. Bull., 2, 1980, p. 607

261
IMMOBILISATION OF ANTIBODIES ON NYLON FOR USE IN ENZYME-LINKED IM-
MUNOASSAY
Hendry, R.M. & Herrman, J.E.
J. Immunol Methods, 35, 1980, p. 285
IMMUNOGLOBULIN G; IMMUNOGLOBULIN E

Amide Amine Polymer

262
EFFECTS OF A SERIES OF NEW SYNTHETIC HIGH POLYMERS ON CANCER
METASTASES
Ferruti, P. et al.
J. Med. Chem., 16, 1973, p. 496

Amide Ester Polymer

263
SYNTHESIS OF COPOLYAMIDE-ESTERS AND SOME ASPECTS INVOLVED IN THEIR
HYDROLYSIS BY LIPASE
Tokiwa, Y. et al.
J. Appl. Polym. Sci., 24, 1979, p. 1701

Caproamide Polymer and Dodecanamide Polymer

264
SPECIFICITY OF POLYMER DEGRADATION IN THE LIVING BODY
Moiseev, Yu.V. et al.
J. Polym. Sci. Polym. Symp., No. 66, 1979, p. 269

Hexamethylenediamine Polymer

265
SEMIPERMEABLE MICROCAPSULES CONTAINING CATALASE FOR ENZYME
REPLACEMENT IN ACATALASAEMIC MICE
Chang, T.M.S. & Poznansky, M.J.
Nature, 218, Apr. 20 1968, p. 243

266
IN VIVO EFFECTS OF SEMIPERMEABLE MICROCAPSULES CONTAINING L-
ASPARAGINASE ON 6C3HED LYMPHOSARCOMA
Chang, T.M.S.
Nature, 229, Jan. 8 1971, p. 117

Phthalamide Polymer

267
STUDIES ON MICROCAPSULES. II. PREPARATION OF POLYPHTHALAMIDE
MICROCAPSULES
Koishi, M. et al.
Chem. Pharm. Bull., 17, 1969, p. 804

BENZOTRIAZOLIDE POLYMER

268
SYNTHESIS AND EXCHANGE REACTIONS OF SOME POLYMERIC BEN-
ZOTRIAZOLIDES
Ferruti, P. et al.
J. Polym. Sci. Polym. Chem., 16, 1978, p. 1435
BUTYLAMINE; MORPHOLINE

BUTADIENE POLYMER

269
SYNTHESIS AND PROPERTIES OF A NEW CLASS OF POTENTIAL BIOMEDICAL
POLYMERS
Rembaum, A. et al.
Biomedical Polymers, 1971
Rembaum, A. & Shen, M. eds.
HEPARIN

270
POLYISOPRENE AND POLYBUTADIENE DERIVATIVES OF TESTOSTERONE
Pinazzi, A. et al.
J. Polym. Sci. Polym. Lett., 12, 1974, p. 447

271
POLYISOPRENE AND POLYBUTADIENE DERIVATIVES OF POTENTIAL BIOMEDICAL
INTEREST
Pinazzi, C.P. et al.
J. Polym. Sci. Polym. Symp., No. 52, 1975, p. 1
QUININE; CHOLESTEROL; TESTOSTERONE

CAPROLACTONE POLYMER

272
UNIFIED MATHEMATICAL MODEL FOR DIFFUSION FROM DRUG-POLYMER COM-
POSITE TABLETS
Fu, J.C. et al.
J. Biomed. Mater. Res., 10, 1976, p. 743
HYDROCORTISONE

273
SUSTAINED DRUG DELIVERY SYSTEMS. I. PERMEABILITY OF POLYCAPROLAC-
TONE, POLYLACTIC ACID AND THEIR COPOLYMERS
Pitt, C.G. et al.
J. Biomed. Mater. Res., 13, 1979, p. 497
PROGESTERONE; TESTOSTERONE; NORGESTREL; NORETHINDRONE; ETHYNYL
ESTRADIOL; IMPLANT

274
SUSTAINED DRUG DELIVERY SYSTEMS. II. FACTORS AFFECTING RELEASE RATES
FROM POLYCAPROLACTONE AND RELATED BIODEGRADABLE POLYESTERS
Pitt, C.G. et al.
J. Pharm. Sci., 68, 1979, p. 1534

CARBONATE POLYMER

275
SUSTAINED RELEASE HORMONAL PREPARATIONS. I. DIFFUSION OF VARIOUS
STEROIDS THROUGH POLYMER MEMBRANES
Kincl, F.A. et al.
Steroids, 11, 1968, p. 673
NORPROGESTERONE; PROGESTERONE; TESTOSTERONE; NORETHINDRONE;
ESTRADIOL; MESTRANOL; CORTICOSTERONE; CORTISOL

276
SURFACE BONDED HEPARIN
Falb, R.D.
Polym. Sci. Technol., 8, 1974, p. 77
Johnson and Johnson Symposium, N.J., July 1974

277
SYNTHESIS AND SOME PROPERTIES OF ANTITHROMBOGENIC POLYMERS
Plate, N.A.
Polymeric Drugs, 1978, p. 63; Academic Press
Donaruma, L.G. & Vogl, O. eds.

278
SPECIFICITY OF POLYMER DEGRADATION IN THE LIVING BODY
Moiseev, Yu.V. et al.
J. Polym. Sci. Polym. Symp., No. 66, 1979, p. 269

CARBOXYLATE POLYMER

279
EFFECTIVE ANTIVIRAL THERAPY OF TWO MURINE LEUKEMIAS WITH AN
INTERFERON-INDUCING SYNTHETIC CARBOXYLATE COPOLYMER
Chirigos, M.A. et al.
Int. J. Cancer, 4, 1969, p. 267

CARBOXYLIC ACID POLYMER

280
INTERACTION OF AMINE DRUGS WITH A POLYCARBOXYLIC ACID ION-
EXCHANGE RESIN
Borodkin, S. & Yunker, M.H.
J. Pharm. Sci., 59, 1970, p. 481
PHENYLPROPANOLAMINE; EPHEDRINE; PSEUDOEPHEDRINE;
DESOXYEPHEDRINE; CARBINOXAMINE MALEATE; DEXTROMETHORPHAN;
CHROMONAR HCl; METHAPYRILENE; QUINIDINE SULPHATE; NEOSTIGMINE
BROMIDE; THIAMINE MONONITRATE

281
POLYACETAL CARBOXYLIC ACIDS: A NEW GROUP OF ANTIVIRAL POLYANIONS
Claes, P. et al.
J. Virol., 5, 1970, p. 313

282
MOLECULAR SCALE DRUG ENTRAPMENT AS A PRECISE METHOD OF CON-
TROLLED DRUG RELEASE. II. FACILITATED DRUG ENTRAPMENT TO POLYMERIC
COLLOIDAL DISPERSIONS
Rhodes, C.T. et al.
J. Pharm. Sci., 59, 1970, p. 1578
CHLORPHENIRAMINE

283
MOLECULAR SCALE DRUG ENTRAPMENT AS A PRECISE METHOD OF CON-
TROLLED DRUG RELEASE. III. IN VITRO AND IN VIVO STUDIES OF DRUG RELEASE
Rhodes, C.T. et al.
J. Pharm. Sci., 59, 1970, p. 1581
PHENYLEPHRINE; PHENYLPROPANOLAMINE

284
BIOLOGICAL ACTIVITY OF POLYCARBOXYLIC ACID POLYMERS
Ottenbrite, R.M. et al.
Polymeric Drugs, 1978, p. 263; Academic Press
Donaruma, L.G. & Vogl, O. eds.

285
STRUCTURE AND BIOLOGICAL ACTIVITIES OF SOME POLYANIONIC POLYMERS
Ottenbrite, R.M.
Anionic Polymeric Drugs, 1980, p. 21
Donaruma, L.G. et al. eds.; Wiley-Interscience

CYANOACRYLATE POLYMERS

Cyanoacrylate Polymer

286
POLYCYANOACRYLATE NANOCAPSULES AS POTENTIAL LYSOSOMOTROPIC
CARRIERS: PREPARATION, MORPHOLOGY AND SORPTIVE PROPERTIES
Couvreur, P. et al.
J. Pharm. Pharmacol., 31, 1979, p. 331

287
ADSORPTION OF ANTINEOPLASTIC DRUGS TO POLYALKYLCYANOACRYLATE
NANOPARTICLES AND THEIR RELEASE IN CALF SERUM
Couvreur, P. et al.
J. Pharm. Sci., 68, 1979, p. 1521

288
TISSUE DISTRIBUTION OF ANTITUMOUR DRUGS ASSOCIATED WITH
POLYALKYLCYANOACRYLATE NANOPARTICLES
Couvreur, P. et al.
J. Pharm. Sci., 69, 1980, p. 199
DACTINOMYCIN; VINBLASTINE

Methyl Cyanoacrylate Polymer

289
IN VIVO METABOLIC DEGRADATION OF POLYMETHYLCYANOACRYLATES VIA
THIOCYANATE
Kulkarni, R.K. et al.
J. Biomed. Mat. Res., 1, 1967, p. 11

DIMETHYL SILOXANE POLYMERS — SEE SILICONE RUBBER

EPICHLOROHYDRIN POLYMER

290
POLYMERIC DRUGS IN THE CHEMOTHERAPY OF MICROBIAL INFECTIONS
Samour, C.M.
Polymeric Drugs, 1978, p. 161; Academic Press
Donaruma, L.G. & Vogl, O. eds.

EPOXY RESINS

291

EPOXY RESIN BEADS AS A PHARMACEUTICAL DOSAGE FORM. I: METHOD OF PREPARATION
Khanna, S.C. & Speiser, P.
J. Pharm. Sci., 58, 1969, p. 1114
CHLORAMPHENICOL; BARIUM SULPHATE; DEHYDROEMETINE DIHYDROCHLORIDE

292

EPOXY RESIN BEADS AS A PHARMACEUTICAL DOSAGE FORM. II: DISSOLUTION STUDIES OF EPOXY-AMINE BEADS AND RELEASE OF DRUG
Khanna, S.C. et al.
J. Pharm. Sci., 58, 1969, p. 1385
CHLORAMPHENICOL

293

HOT EXTRUDED DOSAGE FORMS. I: TECHNOLOGY AND DISSOLUTION KINETICS OF POLYMERIC MATRICES
El-Egakey, M.A. et al.
Pharm. Acta Helv., 46, 1971, p. 31

294

DIFFUSION OF PYRIMETHAMINE FROM SILICONE RUBBER AND FLEXIBLE EPOXY DRUG CAPSULES
Fu, J.C. et al
J. Biomed. Mat. Res., 7, 1973, p. 193

ESTER POLYMERS

Ester Polymer

295
SUSTAINED RELEASE HORMONAL PREPARATIONS. I. DIFFUSION OF VARIOUS
STEROIDS THROUGH POLYMER MEMBRANES
Kincl, F.A. et al.
Steroids, 11, 1968, p. 673
NORPROGESTERONE; PROGESTERONE; TESTOSTERONE; NORETHINDRONE;
ESTRADIOL; MESTRANOL; CORTICOSTERONE; CORTISOL

296
SURFACE BONDED HEPARIN
Falb, R.D.
Polym. Sci. Technol., 8, 1974, p. 77
Johnson and Johnson Symposium, N.J., July 1974

297
PHARMACEUTICAL AND PHARMOKINETIC PRINCIPLES IN THE FORMULATION
OF A SUSTAINED-RELEASE PROCAINAMIDE TABLET
Dahl, S.G. et al.
Pharm. Acta Helv., 51, 1976, p. 204

298
SLOW RELEASE OF WATER SOLUBLE SALTS FROM POLYMERS
Narkis, N. & Narkis, M.
J. Appl. Polym. Sci., 20, 1976, p. 3431
SODIUM CHLORIDE; SODIUM FLUORIDE

299
BIODEGRADABLE POLYMERS FOR SUSTAINED DRUG DELIVERY
Schindler, A. et al.
Contemporary Topics in Polymer Science, 2, 1976, p. 251

Ester Urea Polymer

300
BIODEGRADABLE POLYMERS: CHYMOTRYPTIC DEGRADATION OF LOW
MOLECULAR WEIGHT POLY(ESTER-UREA) CONTAINING PHENYLALANINE
Huang, S.J. et al.
J. Appl. Polym. Sci., 23, 1979, p. 429

Ethylene Sulphonate Polymer

301
HEPARIN-LIKE ACTIVITY OF POLYETHYLENE SULPHONATE
Duncan, C.H. et al.
J. Lab. & Clin. Med., 52, 1958, p. 809
ANTICOAGULANT

302
SODIUM POLYETHYLENE SULPHONATE EFFECTS ON SERUM LIPIDS AND BLOOD
COAGULATION
Kuo, P.T. et al.
Circulation Res., 6, 1958, p. 178

303
ANIONIC POLYELECTROLYTE, POLYETHYLENE SULPHONATE AS A NEW ANTI-
NEOPLASTIC AGENT
Regelson, W. & Holland, J.F.
Nature, 181, 4 Jan. 1958, p. 46

304
EFFECT OF AN ANIONIC POLYELECTROLYTE (POLYETHYLENE SULPHONATE) IN
PATIENTS WITH CANCER
Regelson, W. & Holland, J.F.
Clin. Pharmacol. Ther., 3, 1962, p. 730

305
BIOLOGICAL ACTIVITY OF POLYANIONS: PAST HISTORY AND NEW PROSPEC-
TIVES
Regelson, W.
J. Polym. Sci. Polym. Symp., No. 66, 1979, p. 483

Ethylene Terephthalate Polymer

306
SPECIFICITY OF POLYMER DEGRADATION IN THE LIVING BODY
Moiseev, Yu.V. et al.
J. Polym. Sci. Polym. Symp., No. 66, 1979, p. 269

307
KINETIC SPECIFICITY OF POLYETHYLENE TEREPHTHALATE DEGRADATION IN
THE LIVING BODY
Rudakova, T.E. et al.
J. Polym. Sci. Polym. Symp., No. 66, 1979, p. 277

ETHYLENE POLYMERS AND COPOLYMERS

Ethylene Polymer

308
FURTHER STUDIES OF POLYMERS AS CARCINOGENIC AGENTS IN ANIMALS
Oppenheimer, B.S. et al.
Cancer Res., 15, 1955, p. 333

309
COMPARATIVE STUDY OF SARCOMA FORMATION BY IMPLANTED
POLYETHYLENE FILM AND MESH IN WHITE RATS
Shulmar, J. et al.
Brit. J. Plast. Surg., 16, 1963, p. 336

310
INTRAUTERINE CONTRACEPTION: A NEW APPROACH
Margulis, L.C.
Obstet. Gynecol., 24, 1964, p. 515

311
INVESTIGATION OF FACTORS INFLUENCING RELEASE OF SOLID DRUG DISPER-
SED IN INERT MATRICES
Desai, S.J. et al.
J. Pharm. Sci., 54, 1965, p. 1459

312
TISSUE REACTION INDUCED IN GUINEA PIGS BY PARTICULATE POLYMETHYL
METHACRYLATE, POLYTHENE AND NYLON OF THE SAME SIZE RANGE
Stinson, N.E.
Brit. J. Exp. Pathol., 46, 1965, p. 135

313
INVESTIGATION OF FACTORS INFLUENCING RELEASE OF SOLID DRUG DISPER-
SED IN INERT MATRICES. II. QUANTITATIVE PROCEDURES
Desai, S.J. et al.
J. Pharm. Sci., 55, 1966, p. 1224

314
INVESTIGATION OF FACTORS INFLUENCING RELEASE OF SOLID DRUG DISPERSED IN INERT MATRICES. III.
Desai, S.J. et al.
J. Pharm. Sci., 55, 1966, p. 1230
SULPHANILAMIDE; CAFFEINE; POTASSIUM ACID PHTHALATE

315
KINETICS OF EFFECT LEVELS AFTER HYOSCYAMINE IN SUSTAINED RELEASE TABLETS
Brandstrom, A. & Sjogren, J.
Acta Pharm. Suec., 4, 1967, p. 157

316
DIFFUSION, PERMEATION, AND SOLUBILITY OF SELECTED AGENTS IN AND THROUGH POLYETHYLENE
Gonzales, M.A. et al.
J. Pharm. Sci., 56, 1967, p. 1288
BENZYL ALCOHOL; BENZOIC ACID; BENZALDEHYDE; ACETOPHENONE; METHYLBENZALDEHYDE; METHYL ACETOPHENONE

317
RELEASE RATES OF SOLID DRUG MIXTURES DISPERSED IN INERT MATRICES. I. NONINTERACTING DRUG MIXTURES
Singh, P. et al.
J. Pharm. Sci., 56, 1967, p. 1542
SALICYLIC ACID; BENZOIC ACID

318
RELEASE RATES OF SOLID DRUG. MIXTURES DISPERSED IN INERT MATRICES. II. MUTUALLY INTERACTING DRUG MIXTURES
Singh, P. et al.
J. Pharm. Sci., 56, 1967, p. 1548
BENZOCAINE; CAFFEINE

319
LITHIUM ABSORPTION FROM SUSTAINED RELEASE TABLETS
Andisen, A. & Sjogren, J.
Acta Pharm. Suec., 5, 1968, p. 465
LITHIUM SULPHATE

320
SUSTAINED RELEASE HORMONAL PREPARATIONS. I. DIFFUSION OF VARIOUS
STEROIDS THROUGH POLYMER MEMBRANES
Kincl, F.A. et al.
Steroids, 11, 1968, p. 673
NORPROGESTERONE; PROGESTERONE; TESTOSTERONE; NORETHINDRONE;
ESTRADIOL; MESTRANOL; CORTICOSTERONE; CORTISOL

321
INTERFERON-STIMULATING AND IN VIVO ANTIVIRAL EFFECTS OF VARIOUS
SYNTHETIC ANIONIC POLYMERS
Merigan, T.C. & Finkelstein, M.S.
Virology, 35, 1968, p. 363

322
ENDOMETRIAL MORPHOLOGY AND POLYETHYLENE INTRAUTERINE DEVICES: A
STUDY OF 200 ENDOMETRIAL BIOPSIES
Ober, W.B. et al.
Obstet. Gynecol., 32, 1968, p. 782

323
POLYETHYLENE INTRAUTERINE CONTRACEPTIVE DEVICE: ENDOMETRIAL
CHANGES FOLLOWING LONG TERM USE
Ober, W.B. et al.
J. Amer. Med. Assn., 212, 1970, p. 765

324
BIOLOGICAL ACTIVITY OF POLYMERS
Conning, D.M.
Plastics in Medicine & Surgery
Symposium 15 & 16 Sept. 1971

325
POTASSIUM ABSORPTION FROM SUSTAINED RELEASE TABLETS
Graffner, C. & Sjogren, J.
Acta Pharm. Suec., 8, 1971, p. 13

326
SIDE EFFECTS OF POTASSIUM CHLORIDE IN PRODUCTS WITH DIFFERENT
DISSOLUTION RATES
Graffner, C. & Sjogren, J.
Acta Pharm. Suec., 8, 1971, p. 19

327
ABSORPTION OF ALPRENOLOL IN MAN FROM TABLETS WITH DIFFERENT RATES
OF RELEASE
Johansson, R. et al.
Acta Pharm. Suec., 8, 1971, p. 59

328
EFFECTS OF POTASSIUM TABLETS OF DIFFERENT DISSOLUTION RATES ON
MOTILITY PATTERN OF SMALL INTESTINE IN THE DOG
Sundell, G.
Acta Pharm. Suec., 8, 1971, p. 73
POTASSIUM CHLORIDE

329
RELEASE OF PROGESTERONE FROM POLYETHYLENE DEVICES IN VITRO AND IN
EXPERIMENTAL ANIMALS
Kalkwarf, D.R. et al.
Contraception, 6, 1972, p. 423
MATRIX

330
IN VITRO AND IN VIVO EVALUATION OF SUSTAINED RELEASE TABLETS CON-
TAINING NOREPHEDRINE CHLORIDE
Lindberg, N.O. & Persson, C.G.A.
Acta Pharm. Suec., 9, 1972, p. 237

331
PERMEATION OF AROMATIC COMPOUNDS FROM AQUEOUS SOLUTIONS
THROUGH POLYETHYLENE
Nasim, K. et al.
J. Pharm. Sci., 61, 1972, p. 1775

332
EFFECTS OF STRUCTURE ON PERMEABILITY OF SUBSTITUTED ANILINES FROM
AQUEOUS SOLUTIONS THROUGH POLYETHYLENE
Serota, D.G. et al.
J. Pharm. Sci., 61, 1972, p. 416

333
SURFACE BONDED HEPARIN
Falb, R.D.
Polym. Sci. Technol., 8, 1974, p. 77
Johnson and Johnson Symposium, N.J., July, 1974

334
COMPARATIVE ABSORPTION AND PHARMOKINETIC STUDIES ON PROXYPHYLLINE IN ORDINARY AND SLOW RELEASE TABLETS
Graffner, C. et al.
Acta Pharm. Suec., 11, 1974, p. 125

335
UTERINE PROGESTERONE SYSTEM
Pharriss, B.B.
Intrauterine Devices: Development, Evaluation and Program Implementation, 1974, p. 203; Academic Press
Wheeler, R.G. et al. eds.

336
FERTILITY CONTROL BY INTRAUTERINE RELEASE OF PROGESTERONE
Scommegna, A. et al.
Obstet. Gynecol., 43, 1974, p. 769

337
CONTROLLED RELEASE OF BIOLOGICALLY ACTIVE AGENTS
Yolles, S.
Polym. Sci. Technol., 8, 1974, p. 245
NALOXONE; NALTREXONE; PROGESTERONE; CYTOXAN; DICHLORO-DIAMMINEPLATINUM; MATRIX FILM

338
ANTIMICROBIAL POLYMERS
Ackart, W.B. et al.
J. Biomed. Mater. Res., 9, 1975, p. 55
BENZALKONIUM SALTS; 8-HYDROXYQUINOLINIUM

339
COMPOSITE MEMBRANE ESTRADIOL IMPLANT
Bloch, R. et al.
J. Pharm. Sci., 64, 1975, p. 832

340
SOLUTION-SOLUBILITY DEPENDENCY OF CONTROLLED RELEASE OF DRUG FROM POLYMER MATRIX: MATHEMATICAL ANALYSIS
Chien, Y.W. et al.
J. Pharm. Sci., 64, 1975, p. 1643
ETHYNODIOL DIACETATE

341
THERMODYNAMIC METHOD OF PREDICTING THE TRANSPORT OF STEROIDS IN POLYMER MATRICES
Michaels, A.S. et al.
Amer. Inst. Chem. Eng. J., 21, 1975, p. 1073
PREGNADIENE METHYLDIONE; PROGESTERONE; NORPROGESTERONE; TESTOSTERONE; PREGNATRIENE METHYL ETHINYLACETATE; ESTRADIOL; NORGESTREL; NORETHINDRONE; CORTISOL; PREDNISOLONE; ESTRIOL

342
SLOW RELEASE OF WATER SOLUBLE SALTS FROM POLYMERS
Narkis, N. & Narkis, M.
J. Appl. Polym. Sci., 20, 1976, p. 3431
SODIUM CHLORIDE; SODIUM FLUORIDE

343
PROGRAMMED DIFFUSIONAL RELEASE RATE FROM ENCAPSULATED COSOLVENT SYSTEM
Theeuwes, F. et al.
J. Pharm. Sci., 65, 1976, p. 648
PROGESTERONE

344
DEVELOPMENT OF AN ESTRIOL-RELEASING INTRAUTERINE DEVICE
Baker, R.W. et al.
ACS Coatings and Plastics Preprints, 37, 1977, p. 421

345
SYNTHESIS AND SOME PROPERTIES OF ANTITHROMBOGENIC POLYMERS
Plate, N.A.
Polymeric Drugs, 1978, p. 63, Academic Press
Donaruma, L.G. & Vogl, O. eds.

346
SPECIFICITY OF POLYMER DEGRADATION IN THE LIVING BODY
Moiseev, Yu.V. et al.
J. Polym. Sci. Polym. Symp., No. 66, 1979, p. 269

347
COMPARISON OF SEGMENTED POLYETHER URETHANE WITH POLYETHYLENE IUDs IN RABBITS
Stolzenberg, S.J. et al.
Contraception, 20, 1979, p. 91

348
ECTOPIC PREGNANCIES ASSOCIATED WITH LOW DOSE PROGESTAGEN-
RELEASING IUDs
Diaz, S. et al.
Contraception, 22, 1980, p. 259
MEGESTROL ACETATE; LEVONORGESTREL; NORETHINDRONE

349
IMMOBILISATION OF ENZYMES FOR MEDICAL USES ON PLASTIC SURFACES BY
RADIATION-INDUCED POLYMERISATION AT LOW TEMPERATURES
Kaetsu, I. et al.
J. Biomed. Mat. Res., 14, 1980, p. 199
GLUCOSE PEROXIDASE; UROKINASE

350
METHODOLOGICAL· ASSESSMENTS OF SUBCUTANEOUS IMPLANTATION
TECHNIQUES
Marion, L. et al.
J. Biomed. Mater. Res., 14, 1980, p. 343

351
EFFECT OF SELF-ASSOCIATION OF PHENOL ON ITS TRANSPORT ACROSS
POLYETHYLENE FILM
Mikkelson, T.J. et al.
J. Pharm. Sci., 69, 1980, p. 133

352
CONTROLLED RELEASE MICROPUMP FOR INSULIN DELIVERY AT VARIABLE RATES
Sefton, M.V. & Burns, K.J.
ACS Polym. Preprints, 21, 1980, p. 105

Ethylene-Acrylic Acid Copolymer

353
ANTIMICROBIAL POLYMERS
Ackart, W.B. et al.
J. Biomed. Mater. Res., 9, 1975, p. 55
BENZALKONIUM SALTS; 8-HYDROXYQUINOLINIUM

354
USE OF GRAFT COPOLYMERS AS ENZYME SUPPORTS. THE PREPARATION AND
USE OF POLYETHYLENE-CO-ACRYLIC ACID SUPPORTS
Beddows, C.G. et al.
Polym. Bull., 3, 1980, p. 645
BOVINE SERUM ALBUMIN; PHENYL PROPYLAMINE

Ethylene Glycol Polymer

355
STUDY OF POSSIBLE FORMATION BETWEEN MACROMOLECULES AND CERTAIN PHARMACEUTICALS
Guttman, D.E. & Higuchi, T.
J. Amer. Pharm. Assoc. Sci. Ed., 44, 1955, p. 668
IODINE

356
POSSIBLE COMPLEX FORMATION BETWEEN MACROMOLECULES AND CERTAIN PHARMACEUTICALS. X.
Guttman, D. & Higuchi, T.
J. Am. Pharm. Assn., 45, 1956, p. 659
PHENOL; RESORCINOL; CATECHOL; HYDROQUINONE; TANNIC ACID; PYROGALLOL; BENZOIC ACID; HYDROXYBENZOIC ACID; SALICYLIC ACID; SODIUM SALICYLATE; METHYLPARABEN; BETA NAPHTHOL; PICRIC ACID

357
INTERACTION OF PRESERVATIVES WITH MACROMOLECULES. III. PARAHYDROXYBENZOIC ACID ESTERS IN THE PRESENCE OF SOME HYDROPHILIC POLYMERS
Miyawaki, G.M. et al.
J. Am. Pharm. Assn., 48, 1959, p. 315
PARABENS

358
INTERACTION OF SOME PHARMACEUTICALS WITH MACROMOLECULES. I. EFFECT OF TEMPERATURE ON THE BINDING OF PARABENS AND PHENOLS BY POLYSORBATE 80 AND POLYETHYLENE GLYCOL 4000
Patel, N.K. & Foss, N.E.
J. Pharm. Sci., 53, 1964, p. 94

359
INTERACTION OF LOW MOLECULAR WEIGHT POLYETHYLENE GLYCOLS WITH SORBITOL SOLUTION
Robinson, J. et al.
J. Pharm. Sci., 53, 1964, p. 1245

360
EFFECT OF INERT TABLET INGREDIENTS ON DRUG ABSORPTION. I. EFFECT OF POLYETHYLENE GLYCOL 4000 ON THE INTESTINAL ABSORPTION OF FOUR BARBITURATES
Singh, P. et al.
J. Pharm. Sci., 55, 1966, p. 63
PHENOBARBITAL (2); BARBITANIC ACID; BARBITAL

361
DETERMINATION OF THE DEGREE OF CRYSTALLINITY IN SOLID-SOLID EQUILIBRIA
Allen, D.J. & Kwan, K.C.
J. Pharm. Sci., 58, 1969, p. 1190
INDOMETHACIN

362
PREPARATION AND DISSOLUTION CHARACTERISTICS OF SEVERAL FAST-RELEASE SOLID DISPERSIONS OF GRISEOFULVIN
Chiou, W.L. & Riegelman, S.
J. Pharm. Sci., 58, 1969, p. 1505

363
EFFECT OF POLYETHYLENE GLYCOL 300 ON THE VIABILITY OF BACTERIAL SPORES
Robinson, R.L. & Weinswig, M.H.
J. Pharm. Sci., 58, 1969, p. 275

364
ORAL ABSORPTION OF GRISEOFULVIN IN DOGS: INCREASED ABSORPTION VIA SOLID DISPERSION IN POLYETHYLENE GLYCOL 6000
Chiou, W.L. & Riegelman, S.
J. Pharm. Sci., 59, 1970, p. 937

365
PREPARATION AND TESTING OF A SUSTAINED RELEASE CODEINE PREPARATION
Cordes, G.
Drugs made in Germany, 13, 1970, p. 92
PELLET

366
MOLECULAR SCALE DRUG ENTRAPMENT AS A PRECISE METHOD OF CON-TROLLED DRUG RELEASE. I. ENTRAPMENT OF CATIONIC DRUGS BY POLYMERIC FLOCCULATION
Goodman, H. & Banker, G.S.
J. Pharm. Sci., 59, 1970, p. 1131
METHAPYRILENE

367
SOLUBILITY OF ALKYL BENZOATES. I. EFFECT OF SOME ALKYL p-HYDROXYBENZOATES (PARABENS) ON THE SOLUBILITY OF BENZYL p-HYDROXY BENZOATE
Shihab, F. et al.
J. Pharm. Sci., 59, 1970, p. 1574

368
SOLID DISPERSION APPROACH TO THE FORMULATION OF ORGANIC LIQUID
DRUGS USING POLYETHYLENE GLYCOL 6000 AS A CARRIER
Chiou, W.L. & Smith, L.D.
J. Pharm. Sci., 60, 1971, p. 125
BENZONATATE; CLORFIBRATE; METHYL SALICYLATE; BENZYL BENZOATE;
TOCOPHERYL ACETATE

369
ABSORPTION CHARACTERISTICS OF SOLID DISPERSED AND MICRONISED
GRISEOFULVIN IN MAN
Chiou, W.L. & Riegelman, S.
J. Pharm. Sci., 60, 1971, p. 1377

370
INCREASED DISSOLUTION RATES OF WATER-INSOLUBLE CARDIAC GLYCOSIDES
AND STEROIDS VIA SOLID DISPERSIONS IN POLYETHYLENE GLYCOL 6000
Chiou, W.L. & Riegelman, S.
J. Pharm. Sci., 60, 1971, p. 1569
HYDROCORTISONE ACETATE; TESTOSTERONE; PREDNISOLONE; DIGITOXIN

371
EFFECTS OF NON IONIC SURFACTANTS THAT MODIFY TUBERCULOSIS ON
LIPASE ACTIVITY OF MACROPHAGES
Hart, P.D. & Payne, S.N.
Br. J. Pharmac., 43, 1971, p. 190

372
EFFECTS OF POTASSIUM TABLETS OF DIFFERENT DISSOLUTION RATES ON
MOTILITY PATTERN OF SMALL INTESTINE IN THE DOG
Sundell, G.
Acta Pharm. Suec., 8, 1971, p. 73
POTASSIUM CHLORIDE

373
DECOMPOSITION OF ASPIRIN IN POLYETHYLENE GLYCOLS
Jun, H.W. et al.
J. Pharm. Sci., 61, 1972, p. 1160

374
MEASUREMENT OF DISSOLUTION RATES OF POTASSIUM CHLORIDE FROM
VARIOUS SLOW RELEASE POTASSIUM CHLORIDE TABLETS USING A SPECIFIC ION
ELECTRODE
Thomas, W.H.
J. Pharm. Pharmacol., 25, 1973, p. 27

375
POLYETHYLENE GLYCOL DERIVATIVES OF PROCAINE
Weiner, B.Z. & Zilkha, A.
J. Med. Chem. 16, 1973, p. 573

376
PROSTAGLANDIN DELIVERY BY CERVICAL DILATOR
Balin, H. et al.
J. Reprod. Med., 13, 1974, p. 208
PROSTAGLANDIN E2; PROSTAGLANDIN F2-ALPHA

377
STABILITY OF ASPIRIN IN LIQUID AND SEMI SOLID BASES. IV. POLYETHYLENE
GLYCOL 400 DIACETATE AND TRIETHYLENE GLYCOL DIACETATE
Whitworth, C.W. & Asker, A.F.
J. Pharm. Sci., 63, 1974, p. 1790

378
SOLUTION-SOLUBILITY DEPENDENCY OF CONTROLLED RELEASE OF DRUG
FROM POLYMER MATRIX: MATHEMATICAL ANALYSIS
Chien, Y.W. et al.
J. Pharm. Sci., 64, 1975, p. 1643

379
CORRELATIONS BETWEEN PHYSICAL AND DRUG RELEASE CHARACTERISTICS OF
POLYETHYLENE GLYCOL SUPPOSITORIES
Kellaway, I.W. & Marriott, C.
J. Pharm. Sci., 64, 1975, p. 1162
PREDNISOLONE

380
INFLUENCE OF PEROXIDE IMPURITIES IN POLYETHYLENE GLYCOLS ON DRUG
STABILITY
McGinty, J.W. et al.
J. Pharm. Sci., 64, 1975, p. 356
TRIPELENNAMINE

381
CONTROLLED RELEASE OF TRIPELENNAMINE AND OTHER DRUGS DISPERSED IN
ETHYL CELLULOSE PEG FILMS
Donbrow, M. & Samoelov, Y.
J. Pharm. Pharmacol., 28, 1976, (Suppl.), p. 23P
SALICYLIC ACID; CAFFEINE

382
EFFECT OF POLYETHYLENE GLYCOL 4000 ON DISSOLUTION PROPERTIES OF
SULPHATHIAZOLE POLYMORPHS
Niazi, S.
J. Pharm. Sci., 65, 1976, p. 302

383
APPLICATION OF SOME POLYMERS IN THE PHYSICOCHEMICAL DESIGN OF
TABLET FORMULATION
Salib, N.N. & El-Gamal, S.A.
Die Pharmazie, 31, 1976, p. 718
SULPHANILAMIDE

384
ALTERATION OF IMMUNOLOGICAL PROPERTIES OF BOVINE SERUM ALBUMIN
BY COVALENT ATTACHMENT OF POLYETHYLENE GLYCOL
Abuchowski, A. et al.
J. Biol. Chem., 252, 1977, p. 3578

385
EFFECT OF COVALENT ATTACHMENT OF POLYETHYLENE GLYCOL ON IM-
MUNOGENICITY AND CIRCULATING LIFE OF BOVINE LIVER CATALASE
Abuchowski, A.
J. Biol. Chem., 252, 1977, p. 3582

386
PHARMACEUTICAL APPLICATIONS OF SOLID DISPERSION SYSTEMS: X-RAY DIF-
FRACTION AND AQUEOUS SOLUBILITY STUDIES ON GRISEOFULVIN-PEG 6000
SYSTEM
Chiou, W.L.
J. Pharm. Sci., 66, 1977, p. 989

387
POLYETHYLENE GLYCOL AS A BINDER FOR TABLETS
Shah, R.C. et al.
J. Pharm. Sci., 66, 1977, p. 1551

388
MICROENCAPSULATION AND FORMULATIONS OF SUSTAINED RELEASE CAP-
SULES OF DIAZEPAM
Shukla, A.K. & Sharma, S.N.
Ind. J. Pharm., 39, 1977, p. 100

389
BENZOCAINE DIFFUSION FROM POLYETHYLENE GLYCOL THROUGH HUMAN STRATUM CORNEUM
Belmonte, A.A. & Tsai, W.
J. Pharm. Sci., 67, 1978, p. 517

390
Zn-DTPA ADMINISTERED BY SLOW-RELEASE IMPLANT
Calder, S.E. et al.
Health Phys., 35, 1978, p. 785
DIETHYLENETRIAMINEPENTAACETATE

391
FLUIDISED BED COATING TECHNIQUE FOR PRODUCTION OF SUSTAINED RELEASE GRANULES
Friedman, M. & Donbrow, M.
Drug Dev. Ind. Pharm., 4, 1978, p. 319
SALICYLIC ACID; CAFFEINE

392
MOLECULAR INTERACTION BETWEEN E-PROSTAGLANDINS AND SELECTED POLYMERS AND ITS POTENTIAL UTILISATION IN ORAL DOSAGE FORM DESIGN
Fung, H.L. & Cho, M.J.
J. Pharm. Sci., 67, 1978, p. 971
PROSTAGLANDINS E-SERIES; MATRIX

393
NITROGLYCERIN STABILITY IN POLYETHYLENE GLYCOL 400 AND POVIDONE SOLUTIONS
Suphajettrn, P. et al.
J. Pharm. Sci., 67, 1978, p. 1394

394
SUSTAINED RELEASE OF DRUGS FROM ETHYLCELLULOSE-POLYETHYLENE GLYCOL FILMS AND KINETICS OF DRUG RELEASE
Samuelov, Y. et al.
J. Pharm. Sci., 68, 1979, p. 325
SALICYLIC ACID; TRIPELENNAMINE; CAFFEINE

395
SOLUBLE, NONANTIGENIC POLYETHYLENE GLYCOL-BOUND ENZYMES
Davies, F.F. et al.
Biomedical Polymers: Polymeric Materials and Pharmaceuticals for Biomedical Use
Goldberg, E.P. & Nakajima, A. eds, 1980, p. 441; Academic Press
ADENOSINE DEAMINASE; ARGINASE; CATALASE; GLUTAMINASE-ASPARAGINASE; PHENYLALANINE AMMONIALYASE; SUPEROXIDE DISMUTASE; TRYPSIN; URICASE

396
VEHICLE EFFECTS IN PERCUTANEOUS ABSORPTION: IN VITRO STUDY OF INFLUENCE OF SOLVENT POWER AND MICROSCOPIC VISCOSITY OF VEHICLE ON BENZOCAINE RELEASE FROM SUSPENSION HYDROGELS
DiColo, G. et al.
J. Pharm. Sci., 69, 1980, p. 387

397
SUSTAINED RELEASE CHLORHEXIDINE PREPARATIONS FOR TOPICAL USE
Friedman, M. et al.
J. Dental Res., 59, 1980, p. 905

398
HYDROGELS FOR THE CONTROLLED RELEASE OF PROSTAGLANDIN E2
Graham, N.B. et al.
ACS Polymer Preprints, 21, 1980, p. 104

399
CONTROLLED RELEASE OF MULTI-COMPONENT CYTOTOXIC AGENTS FROM RADIATION POLYMERISED COMPOSITES
Kaetsu, I. et al.
Biomaterials, 1, 1980, p. 17
MITOMYCIN; ADRYAMYCIN; TETRAHYDROFURYL FLUOROURACIL

400
CONTROLLED SLOW RELEASE OF CHEMOTHERAPEUTIC DRUGS FOR CANCER FROM MATRICES PREPARED BY RADIATION POLYMERISATION AT LOW TEMPERATURES
Kaetsu, I.
J. Biomed. Mat. Res., 14, 1980, p. 185
MITOMYCIN; 5-FLUOROURACIL; BLEOMYCIN

401
MICROCOLUMN CHROMATOGRAPHIC CLEAN UP AND GLC DETERMINATION
OF 11-METHYL-16,16-DIMETHYL PROSTAGLANDIN E2 IN POLYETHYLENE
GLYCOL 400 SOLUTIONS
Quadrel, R.F. et al.
J. Pharm. Sci., 69, 1980, p. 718

402
PASSAGE OF MOLECULES THROUGH THE WALL OF THE GASTROINTESTINAL
TRACT. II. APPLICATION OF LOW MOLECULAR WEIGHT POLYETHYLENE GLYCOL
AND A DETERMINISTIC MATHEMATICAL MODEL FOR DETERMINING INTESTINAL
PERMEABILITY IN MAN
Sundqvist, T. et al.
Gut., 21, 1980, p. 208

403
POLYETHYLENE GLYCOLS AS SOLVENTS IN IMPLANTABLE OSMOTIC PUMPS
Will, P.C. et al.
J. Pharm. Sci., 69, 1980, p. 747

Ethylene Imine Polymer

404
POLYMERISATION OF HETEROCYCLES RELATED TO BIOMEDICAL POLYMERS
Kropachev, V.A.
Pure & Appl. Chem., 48, 1976, p. 355
CHEMOTHERAPY

405
RIGID SUPPORTS FOR THE IMMOBILISATION OF ENZYMES
Roger, G.P. et al.
J. Macromol. Sci. Chem., A10, 1976, p. 245
TRYPSIN

406
GI PHARMACOLOGY OF POLYETHYLENE IMINE. I. EFFECTS ON GASTRIC
EMPTYING IN RATS
Melamed, S. et al.
J. Pharm. Sci., 66, 1977, p. 899

407
USE OF MICROCAPSULES AS TIMED-RELEASE PARENTERAL DOSAGE FORM: AP-
PLICATION AS RADIO PHARMACEUTICAL IMAGING AGENT
Scheu, J.D. et al.
J. Pharm. Sci., 66, 1977, p. 172
GOLD SODIUM THIOSULPHATE

408
GI PHARMACOLOGY OF POLYETHYLENE IMINE. II. MOTOR ACTIVITY IN ANAESTHETISED DOGS
Tansy, M.F. et al.
J. Pharm. Sci., 66, 1977, p. 902

409
SYNTHESIS AND CHARACTERISATION OF POLYMERIC DERIVATIVES OF THE AN-TITUMOUR AGENT METHOTREXATE
Przybylski, M. et al.
Makromol. Chem., 179, 1978, p. 1719

410
RECENT PROBLEMS CONCERNING FUNCTIONAL MONOMERS AND POLYMERS CONTAINING NUCLEIC ACID BASES
Takemoto, K.
Polymeric Drugs, 1978, p. 103; Academic Press
Donaruma, L.G. & Vogl, O. eds.

411
CHARACTERISATION OF POLYETHYLENE IMINE MODIFIED WITH ORGANOTIN HALIDES. THERMAL, SOLUBILITY, AND FUNGAL PROPERTIES
Carraher, C.E. et al.
J. Appl. Polym. Sci., 23, 1979, p. 1501

Ethylene Maleic Acid Copolymer

412
DISSOLUTION OF MACROMOLECULES. II. DISSOLUTION OF AN ETHYLENE-MALEIC ACID COPOLYMER
Heyd, A. et al.
J. Pharm. Sci., 59, 1970, p. 947

413
MODIFICATION OF FIBRIN BY SYNTHETIC POLYMERS
Kovacs, G. et al.
J. Polym. Sci. Polym. Symp., No. 66, 1979, p. 201

Ethylene Maleic Anhydride Copolymer (Melethamer)

414
SYNTHETIC POLYELECTROLYTES AS TUMOUR INHIBITORS
Regelson, W. et al.
Nature, 186, No. 4727, 1960, p. 778

415
POLYMERIC PHARMACEUTICAL COATING MATERIALS. I. PREPARATION AND PROPERTIES
Lappas, L.C. & McKeehan, W.
J. Pharm. Sci., 54, 1965, p. 176

416
EXPERIMENTAL PRODUCTION OF DIARRHOEA AND ITS PREVENTION BY MALETHAMER IN MONKEYS
Lin, T.M.
Arch. Int. Pharmacodyn., 169, 1967, p. 147
CHEMOTHERAPY

417
MECHANISM STUDY OF THE ACTION OF MELETHAMER IN STAPHYLOCOCCUS ENTEROTOXIN INDUCED DIARRHOEA IN MONKEYS
Lin, T.M.
Arch. Int. Pharmacodyn., 169, 1967, p. 162
CHEMOTHERAPY

418
INTERFERON-STIMULATING AND IN VIVO ANTIVIRAL EFFECTS OF VARIOUS SYNTHETIC ANIONIC POLYMERS
Merigan, T.C. & Finkelstein, M.S.
Virology, 35, 1968, p. 363
CHEMOTHERAPY

419
DISSOLUTION OF MACROMOLECULES. I. SURFACE PHENOMENA ASSOCIATED WITH POLYMER DISSOLUTION
Heyd, A. et al.
J. Pharm. Sci., 58, 1969, p. 586

420
MACROMOLECULAR DISSOLUTION: TEMPERATURE EFFECTS ON POLYMER-DRUG PREPARATION
Heyd, A.
J. Pharm. Sci., 59, 1970, p. 1526
PHENYLPROPANOLAMINE

421
EFFECT OF MELETHAMER ON THE EXCRETION AND PLASMA LEVELS OF SODIUM, POTASSIUM AND CHLORIDE
Tust, R.H. & Lin, T.M.
Proc. Soc. Exp. Biol. & Med., 135, 1970, p. 72

422
POLYMER-DRUG INTERACTION: STABILITY OF AQUEOUS GELS CONTAINING
NEOMYCIN SULPHATE
Heyd, A.
J. Pharm. Sci., 60, 1971, p. 1343

423
CARRIER-BOUND BIOLOGICALLY ACTIVE SUBSTANCES AND THEIR
APPLICATIONS
Orth, H.D. & Brummer, W.
Angew. Chem. Int. Edn., 11, 1972, p. 249
CHYMOTRYPSIN; KALLIKREIN; PLASMIN; TRYPSIN

424
MODIFICATION OF FIBRIN BY SYNTHETIC POLYMERS
Kovacs, G. et al.
J. Polym. Sci. Polym. Symp., No. 66, 1979, p. 201

Ethylene-Methacrylic Acid Copolymer

425
ANTIMICROBIAL POLYMERS
Ackart, W.B. et al.
J. Biomed. Mater. Res., 9, 1975, p. 55

Ethylene-vinyl Acetate Copolymer (EVA)

426
STRUCTURE AND PROPERTY RELATIONSHIPS IN EVA COPOLYMERS
Salyer, I.O. & Kenyon, A.S.
J. Polym. Sci., A1, 9, 1971, p. 3083

427
EFFECT OF PILOCARPINE OCUSERT WITH DIFFERENT RELEASE RATES ON
OCULAR PRESSURE
Armaly, M.F. & Rao, K.R.
Invest. Ophthalmol, 12, 1973, p. 491

428
PROLONGED RELEASE HYDROCORTISONE THERAPY
Lerman, S, et al.
Can. J. Ophthalmol., 8, 1973, p. 114

429
DRUG DELIVERY SYSTEMS: DESIGN CRITERIA
Cowsar, D.R.
Polym. Sci. Technol., 8, 1974, p. 237
PROGESTERONE; IUD

430
PILOCARPINE OCUSERT: A NEW DRUG DELIVERY SYSTEM
Friederich, R.L.
Ann. Ophthalmol., 6, 1974, p. 1279

431
UTERINE THERAPEUTIC SYSTEM FOR LONG TERM CONTRACEPTION. II.
CLINICAL CORRELATIONS
Martinez-Manautou, J. et al.
Fertil. & Steril., 25, 1974, p. 922
PROGESTERONE

432
PROGESTASERT: A UTERINE THERAPEUTIC SYSTEM FOR LONG-TERM
CONTRACEPTION. I. PHILOSOPHY AND CLINICAL EFFICACY
Pharriss, B.B. et al.
Fertil. & Steril., 25, 1974, p. 915
PROGESTERONE

433
PROGRESS IN THE DEVELOPMENT OF THE PROGESTASERT 65 PROGESTERONE
THERAPEUTIC SYSTEM FOR CONTRACEPTION
Place, V.A. & Pharriss, B.B.
J. Reprod. Med., 13, 1974, p. 66
IUD

434
INTRAUTERINE CONTRACEPTION WITH THE PROGESTERONE-T DEVICE
Rosado, A.
Contraception, 9, 1974, p. 39

435
OCULAR THERAPY BY CONTROLLED DRUG DELIVERY: THE OCUSERT SYSTEM
Shell, J.W.
Ophthalmic Surg., 5, 1974, p. 73
PILOCARPINE

436
EVALUATION OF THE PILOCARPINE OCUSERT
Worthen, D.M. et al.
Invest. Ophthalmology, 13, 1974, p. 296

437
CONTRACEPTION BY INTRAUTERINE RELEASE OF PROGESTERONE: EFFECTS OF
ENDOMETRIAL TRACE ELEMENTS, ENZYMES AND STEROIDS
Hagenfeldt, K. & Landgren, B.M.
J. Steroid Biochem., 6, 1975, p. 895

438
PILOCARPINE OCUSERT SYSTEM FOR SUSTAINED CONTROL OF OCULAR HYPER-
TENSION
Macoul, K.L.
Arch. Ophthalmol., 93, 1975, p. 587

439
CONTRACEPTION BY INTRAUTERINE RELEASE OF PROGESTERONE. CLINICAL
RESULTS
Martinez-Manautou, J.
J. Steroid. Biochem., 6, 1975, p. 889

440
THERMODYNAMIC METHOD OF PREDICTING THE TRANSPORT OF STEROIDS IN
POLYMER MATRICES
Michaels, A.S. et al.
Amer. Inst. Chem. Eng. J., 21, 1975, p. 1073
PREGNADIENE METHYLDIONE; PROGESTERONE; NORPROGESTERONE;
TESTOSTERONE; PREGNATRIENE METHYL ETHINYLACETATE; ESTRADIOL;
NORGESTREL; NORETHINDRONE; CORTISOL; PREDNISOLONE; ESTRIOL

441
COMPARATIVE PHARMACOLOGICAL EFFECTS OF PILOCARPINE ADMINISTERED
TO NORMAL SUBJECTS BY EYEDROPS OR BY OCULAR THERAPEUTIC SYSTEMS
Place, V.A. et al.
Amer. J. Ophthalmol, 80, 1975, p. 706

442
CONTRACEPTION BY INTRAUTERINE RELEASE OF STEROIDS
Zaffaroni, A.
J. Steroid Biochem., 6, 1975, p. 883
PROGESTERONE

443
VISUAL EFFECTS OF PILOCARPINE IN GLAUCOMA
Brown, H.S. et al.
Arch. Ophthalmol., 94, 1976, p. 1716

444
UNIFIED MATHEMATICAL MODEL FOR DIFFUSION FROM DRUG-POLYMER
COMPOSITE TABLETS
Fu, J.C. et al.
J. Biomed. Mater. Res., 10, 1976, p. 743
HYDROCORTISONE

445
EFFECT OF PILOCARPINE OCULAR THERAPEUTIC SYSTEMS ON DIURNAL
CONTROL OF INTRAOCULAR PRESSURE
Fraunfelder, F.T. et al.
Ann. Ophthalmol., 8, 1976, p. 1031

446
SUSTAINED RELEASE OF MACROMOLECULES FROM POLYMERS
Langer, R. & Folkman, J.
Midland Macromol. Inst., Polymer Delivery Systems, 5th Int. Symp., Aug.
1976, p. 175
SOYBEAN TRYPSIN INHIBITOR; LYSOZYME; ALKALINE PHOSPHATASE;
CATALASE; INSULIN; HEPARIN

447
POLYMERS FOR THE SUSTAINED RELEASE OF PROTEINS AND OTHER
MACROMOLECULES
Langer, R. & Folkman, J.
Nature, 263, Oct. 28, 1976, p. 797

448
ISOLATION OF A CARTILAGE FACTOR THAT INHIBITS TUMOUR
NEOVASCULARISATION
Langer, R. et al.
Science, 193, 1976, p. 70

449
ROLE OF POLYMERS IN OPTIMISING THERAPEUTIC EFFECTIVENESS OF DRUGS
Leeper, H. & Benson, H.
SPE 34th Ann. Tech. Conf., April 1976, p. 653
PILOCARPINE; PROGESTERONE

450
CONTROL OF MENSTRUAL BLEEDING DURING HAEMODIALYSIS
Newton, J. et al.
Brit. Med. J., No. 6016, 24th Apr. 1976, p. 1016
PROGESTERONE; IUD

451
CLINICAL EXPERIENCE WITH THE UTERINE PROGESTERONE SYSTEM
PROGESTASERT
Zador, G. et al.
Contraception, 13, 1976, p. 559

452
DEVELOPMENT OF AN ESTRIOL-RELEASING INTRAUTERINE DEVICE
Baker, R.W. et al.
ACS Coatings & Plastics Preprints, 37, 1977, p. 421

453
INHIBITION OF NEOVASCULARISATION BY AN EXTRACT DERIVED FROM
VITREOUS
Brem, S. et al.
Am. J. Ophthalmol., 84, 1977, p. 323

454
EFFECTS OF OCUSERT PILOCARPINE ON ANTERIOR CHAMBER DEPTH, VISUAL
ACUITY AND INTRAOCULAR PRESSURE IN MAN
Drance, S.M. et al.
Canad. J. Ophthalmol., 12, 1977, p. 24

455
USE OF POLYMERS FOR THE SUSTAINED RELEASE OF PROTEINS AND OTHER
LARGE MOLECULES
Langer, R. & Folkman, J.
ACS Polymer Preprints, 18, No. 2, 1977, p. 379
LYSOZYME; ALKALINE PHOSPHATASE

456
ROLE OF POLYMERS IN OPTIMISING THERAPEUTIC EFFECTIVENESS OF DRUGS
Leeper, H. & Benson, H.
Polym. Engng. Sci., 17, 1977, p. 42
PILOCARPINE; PROGESTERONE; CAPSULE; IUD

457
PROGESTASERT AND ECTOPIC PREGNANCY
Snowden, R.
Brit. Med. J., ii, 1977, p. 1600
PROGESTERONE; IUD

458
SEQUENCE OF EVENTS IN THE REGRESSION OF CORNEAL CAPILLARIES
Ausprunk, D.H. et al.
J. Lab. Invest., 38, 1978, p. 284
SODIUM HYDROXIDE

459
MATHEMATICAL MODELS FOR THE RELEASE OF DRUGS FROM MATRIX TABLETS
Cobby, J.
J. Biomed. Mater. Res., 12, 1978, p. 627
HYDROCORTISONE

460
EFFECT OF COMONOMER RATIO ON HYDROCORTISONE DIFFUSION FROM
SUSTAINED-RELEASE COMPOSITE CAPSULES
Fu, J.C. et al.
J. Biomed. Mater. Res., 12, 1978, p. 249

461
NO INCREASE OF THE FIBRINOLYTIC ACTIVITY OF THE HUMAN ENDOMETRIUM
BY PROGESTERONE-RELEASING IUD
Liedholm, P. et al.
Contraception, 17, 1978, p. 531

462
IMMOBILISED ENZYMES
Manecke, G. & Schlunsen, J.
Polymeric Drugs, 1978, p. 39, Academic Press
Donaruma, L.G. & Vogl, O. eds.

463
CLINICAL EXPERIENCE WITH THE INTRAUTERINE PROGESTERONE
CONTRACEPTIVE SYSTEM
Pharriss, B.B.
J. Reprod. Med., 20, 1978, p. 155

464
DIFFERENCES IN EFFECT OF COPPER AND PROGESTASERT IUDs ON
FIBRINOLYTIC ACTIVITY OF THE ENDOMETRIUM IN THE RABBIT
Liedholm, P. & Sjoberg, N.O.
Contraception, 19, 1979, p. 443
PROGESTERONE

465
COLLABORATIVE STUDY OF THE PROGESTERONE INTRAUTERINE DEVICE
PROGESTASERT
Newton, J. et al.
Contraception, 19, 1979, p. 575

466
COMPARATIVE STUDY OF THE EFFECT OF THE PROGESTASERT AND GRAVIGARD
IUDs ON DYSMENORRHOEA
Pizzaro, E. et al.
Contraception, 20, 1979, p. 455

467
SINGLE STEP IMMUNISATION BY SUSTAINED ANTIGEN RELEASE
Preis, I. & Langer, R.S.
J. Immunol. Meth., 28, 1979, p. 193
BOVINE SERUM ALBUMIN; GAMMA GLOBULIN; RIBONUCLEASE

468
KINETICS OF POLYMERIC DELIVERY SYSTEMS FOR PROTEINS AND OTHER
MACROMOLECULES
Rhine, W.D. et al.
ACS Polymer Preprints, 20, No. 1, 1979, p. 596

469
ONE MONTH OF SUSTAINED RELEASE OF INSULIN FROM A POLYMER IMPLANT
Creque, H.M. et al.
Diabetes, 29, 1980, p. 37

470
LEVELS OF HCG IN SUBJECTS USING THE PROGESTASERT INTRAUTERINE
PROGESTERONE SYSTEM
Custo, G.M. et al.
Contraception, 21, 1980, p. 311

471
SELECTED EVENTS FOLLOWING INSERTION OF THE PROGESTASERT SYSTEM
Gibor, Y. & Mitchell, C.
Contraception, 21, 1980, p. 491
PROGESTERONE; IUD

472
ENDOMETRIAL ESTROGEN AND PROGESTIN RECEPTORS IN WOMEN BEARING
A PROGESTERONE-RELEASING INTRAUTERINE DEVICE
Janne, O. & Ylostalo, P.
Contraception, 22, 1980, p. 19

473
SIMPLE METHOD FOR STUDYING CHEMOTAXIS USING SUSTAINED RELEASE OF
ATTRACTANTS FROM INERT POLYMERS
Langer, R. et al.
Can. J. Microbiol., 26, 1980, p. 274

474
POLYMERS FOR SUSTAINED MACROMOLECULE RELEASE: PROCEDURES TO
FABRICATE REPRODUCIBLE DELIVERY SYSTEMS AND CONTROL RELEASE KINETICS
Rhine, W.D. et al.
J. Pharm. Sci., 69, 1980, p. 265
BOVINE SERUM ALBUMIN; BETA-LACTOGLOBULIN; LYSOZYME

475
PROSPECTIVE STUDY OF PLASMA PROLACTIN LEVELS IN WOMEN USING THE
PROGESTERONE RELEASING INTRAUTERINE DEVICE
Spellacy, W.N. & Buhi, W.C.
Contraception, 19, 1979, p. 91

476
EFFECT OF PILOCARPINE OCUSERT ON OCULAR PRESSURE
Armaly, M.F. & Rao, K.R.
Symp. on Ocular Therapy, Vol. 6, St. Louis
Leopold, I.H. ed.

TETRAFLUOROETHYLENE POLYMER

477
FURTHER STUDIES OF POLYMERS AS CARCINOGENIC AGENTS IN ANIMALS
Oppenheimer, B.S. et al.
Cancer Res., 15, 1955, p. 333

478
SUSTAINED RELEASE HORMONAL PREPARATIONS. I. DIFFUSION OF VARIOUS
STEROIDS THROUGH POLYMER MEMBRANES
Kincl., F.A. et al.; Steroids, 11, 1968, p. 673
NORPROGESTERONE; PROGESTERONE; TESTOSTERONE; NORETHINDRONE;
ESTRADIOL; MESTRANOL; CORTICOSTERONE; CORTISOL

479
EXPERIMENTAL ERRORS RESULTING FROM UPTAKE OF LIPOPHILIC DRUGS BY
SOFT PLASTIC MATERIALS
Minder, R. et al.
Biochem. Pharmacol., 19, 1970, p. 2179
IMIPRAMINE

480
SURFACE BONDED HEPARIN
Falb, R.D.
Polym. Sci. Technol., 8, 1974, p. 77
Johnson and Johnson Symposium, N.J., July 1974

481
LONG TERM IMPLANTABLE AQUEOUS DELIVERY SYSTEM FOR THE EXTERNAL
RABBIT EYE
Rehkopf, P.G. et al.
Invest. Ophthalmol. Vis. Sci., 19, 1980, p. 428
SALINE

GALACTURONIC ACID POLYMER

482
EVALUATION OF A PROLONGED ACTION ORAL ANTIHISTAMINIC
PREPARATION AS TREATMENT FOR ALLERGIC DISORDERS
Goldberg, R.I. & Shuman, F.I.
Clin. Med., 72, 1965, p. 1475
PYRILAMINE TANNATE; CHLORPHENIRAMINE TANNATE; AMPHETAMINE
TANNATE

GLUTAMIC ACID POLYMER

483
POLYMERIC DRUGS IN THE CHEMOTHERAPY OF MICROBIAL INFECTIONS
Samour, C.M.
Polymeric Drugs, 1978, p. 161, Academic Press
Donaruma, L.G. & Vogl, O. eds.

GLUTARALDEHYDE POLYMER

484
POLYGLUTARALDEHYDE: A NEW REAGENT FOR COUPLING PROTEINS TO
MICROSPHERES AND FOR LABELLING SURFACE RECEPTORS
Rembaum, A. et al.
J. Immunol. Meth., 24, 1978, p. 239

GLYCOLIC ACID POLYMERS AND COPOLYMERS

Glycolic Acid Polymer

485
BIODEGRADABLE SYSTEMS FOR THE SUSTAINED RELEASE OF FERTILITY-REGULATING AGENTS
Benagiano, G. & Gabelnick, H.L.
J. Steroid. Biochem., 11, 1979, p. 449
NORETHISTERONE; LEVONORGESTREL

486
LACTIC/GLYCOLIC ACID POLYMERS
Wise, D.L. et al.
Drug Carriers in Biology and Medicine, 1979, p. 237
Gregoriadis, G. ed.

487
SUSTAINED RELEASE OF A DUAL ANTIMALARIAL SYSTEM
Wise, D.L. et al.
J. Pharm. Pharmacol., 31, 1979, p. 201
SULPHADIAZINE; DIAMINONAPHTHYL SULPHONYL QUINAZOLINE

Glycolic Acid — Lactic Acid Copolymer

488
SUSTAINED RELEASE OF AN ANTIMALARIAL DRUG USING A COPOLYMER OF GLYCOLIC/LACTIC ACID
Wise, D.L. et al.
Life Sciences, 19, 1976, p. 867
DIAMINONAPHTHYL SULPHONYL QUINAZOLINE (WR-158122)

ISOBUTYLENE-MALEIC ANHYDRIDE COPOLYMER

489
INTERFERON-STIMULATING AND IN VIVO ANTIVIRAL EFFECTS OF VARIOUS SYNTHETIC ANIONIC POLYMERS
Merigan, T.C. & Finkelstein, M.S.
Virology, 35, 1968, p. 363
CHEMOTHERAPY

ISOCYANATE POLYMERS
Diphenyl Methane Diisocyanate Polymer — see Urethane Polymer

ISOPRENE POLYMER

490
ANTITHROMBIN ACTIVITY OF A POLYELECTROLYTE SYNTHESISED FROM CIS-1,4-POLYISOPRENE
Beugeling, T. et al.
J. Biomed. Mater. Res., 8, 1974, p. 375

491
POLYISOPRENE AND POLYBUTADIENE DERIVATIVES OF TESTOSTERONE
Pinazzi, A. et al.
J. Polym. Sci. Polym. Lett., 12, 1974, p. 447

492
POLYISOPRENE AND POLYBUTADIENE DERIVATIVES OF POTENTIAL BIOMEDICAL INTEREST
Pinazzi, C.P. et al.
J. Polym. Sci. Polym. Symp., No. 52, 1975, p. 1
QUININE; CHOLESTEROL; TESTOSTERONE

493
POLYMERS WITH POTENTIAL PHARMACEUTICAL PROPERTIES: FIXATION OF HISTAMINE ON THE MACROMOLECULAR CHAIN OF POLYISOPRENE
Pinazzi, C. et al.
J. Polym. Sci. Polym. Chem., 15, 1977, p. 1319

494
ANTICOAGULANT ACTIVITY OF POLYELECTROLYTES AND THEIR BEHAVIOUR IN SURFACE COATINGS
Sederel, L.C. et al.
Plastics in Medicine and Surgery, June 1979, p. 35.1
Plastics and Rubber Institute

495
SYNTHETIC POLYMERS WITH ANTICOAGULANT ACTIVITY
Van der Does, L. et al.
J. Polym. Sci. Polym. Symp., No. 66, 1979, p. 337

LACTIC ACID POLYMERS AND COPOLYMERS

Lactic Acid Polymer

496
BIODEGRADABLE POLYLACTIC ACID POLYMERS
Kulkarni, R.K. et al.
J. Biomed. Mat. Res., 5, 1971, p. 169

497
LONG-ACTING DELIVERY SYSTEMS FOR NARCOTIC ANTAGONISTS
Woodland, J.H.R. et al.
J. Med. Chem., 16, 1973, p. 897
CYCLAZOCINE

498
INJECTION METHOD FOR DELIVERY OF LONG ACTING NARCOTIC ANTAGONISTS
Leafe, T.D. et al.
Adv. Biochem. Psychopharmacol., 8, 1974, p. 569
CYCLAZOCINE

499
LONG ACTING DELIVERY SYSTEMS FOR NARCOTIC ANTAGONISTS
Yolles, S. et al.
Adv. Expl. Med. & Biol., 47, 1974
Controlled Release of Biologically Active Agents
Tanquary, A.C. & Lacey, R.E. eds.
CYCLAZOCINE; NALOXONE; NALTREXONE

500
CONTROLLED RELEASE OF BIOLOGICALLY ACTIVE AGENTS
Yolles, S.
Polym. Sci. Technol., 8, 1974, p. 245
NALOXONE; NALTREXONE; PROGESTERONE; CYTOXAN; DICHLORODIAM-
MINEPLATINUM

501
TIMED-RELEASE DEPOT FOR ANTICANCER AGENTS
Yolles, S. et al.
J. Pharm. Sci., 64, 1975, p. 115
CYCLOPHOSPHAMIDE; DICHLORODIAMMINEPLATINUM

502
LONG ACTING DELIVERY SYSTEMS FOR NARCOTIC ANTAGONISTS. II. RELEASE
RATES OF NALTREXONE FROM POLYLACTIC ACID COMPOSITES
Yolles, S. et al.
J. Pharm. Sci., 64, 1975, p. 348

503
AN INJECTABLE SUSTAINED RELEASE FERTILITY CONTROL SYSTEM
Anderson, L.C. et al.
Contraception, 13, 1976, p. 375
NORETHISTERONE

504
IN VIVO AND IN VITRO EVALUATION OF A MICROENCAPSULATED NARCOTIC
ANTAGONIST
Mason, N. et al.
J. Pharm. Sci., 65, 1976, p. 847
CYCLAZOCINE

505
TESTING OF DRUG DELIVERY SYSTEMS FOR USE IN THE TREATMENT OF
NARCOTIC ADDICTION
Reuning, R.H. et al.
Res. Monogr. Ser. — Natl. Inst. Drug Abuse (US), 1976, (4), p. 43
NALOXONE; NALTREXONE

506
CHARACTERISATION OF MICROCAPSULES CONTAINING NALTREXONE OR
NALTREXONE PAMOATE
Thies, C.
ACS Coatings & Plastics Preprints, 36, 1976, p. 372
c.f. ACS Symp. Series, 33, 1976, p. 190

507
DEVELOPMENT OF SUSTAINED ACTION PREPARATIONS OF NARCOTIC
ANTAGONISTS
Willette, R.E.
Natl. Inst. Drug Abuse Monogr. Ser., 9, 1976, p. 31
NALTREXONE

508
CONTROLLED RELEASE OF BIOLOGICALLY ACTIVE DRUGS
Yolles, S. et al.
Bull. Parenter. Drug Assoc., 30, 1976, p. 306
PROGESTERONE; ESTRADIOL; CYCLAZOCINE; NALOXONE; NALTREXONE;
CYTOXAN; DICHLORODIAMMINEPLATINUM

509
CONTROLLED RELEASE OF BIOLOGICALLY ACTIVE AGENTS
Yolles, S. et al.
ACS Organic Coatings & Plastics Preprints, 36, 1976, p. 332
c.f. ACS Symp. Series, 33, 1976, p. 123
PROGESTERONE; ESTRADIOL; CYCLAZOCINE; NALTREXONE

510
SUSTAINED RELEASE OF SULPHADIAZINE
Wise, D.L. et al.
J. Pharm. Pharmacol., 30, 1978, p. 686
IMPLANT

511
TIME-RELEASE DEPOT FOR ANTICANCER DRUGS: RELEASE OF DRUGS COVALEN-
TLY BONDED TO POLYMERS
Yolles, S.
J. Parenter Drug Assoc., 32, 1978, p. 188
DICHLORODIAMINEPLATINUM; CYCLOPHOSPHAMIDE; DOXORUBICIN

512
TIMED RELEASE DEPOT FOR ANTICANCER AGENTS. II.
Yolles, S. et al.
Acta Pharm. Suec., 15, 1978, p. 382
DICHLORODIAMINEPLATINUM; CYCLOPHOSPHAMIDE; DOXORUBICIN

513
BIODEGRADABLE SYSTEMS FOR THE SUSTAINED RELEASE OF FERTILITY-
REGULATING AGENTS
Benagiano, G. & Gabelnick, H.L.
J. Steroid Biochem., 11, 1979, p. 449
NORETHISTERONE; LEVONORGESTREL

514
SUSTAINED RELEASE OF A DUAL ANTIMALARIAL SYSTEM
Wise, D.L. et al.
J. Pharm. Pharmacol., 31, 1979, p. 201
SULPHADIAZINE; DIAMINONAPHTHYLSULPHONYL QUINAZOLINE

515
SUSTAINED DRUG DELIVERY SYSTEMS. I. PERMEABILITY OF POLYCAPROLAC-
TONE, POLYLACTIC ACID AND THEIR COPOLYMERS
Pitt, C.G. et al.
J. Biomed. Mater. Res., 13, 1979, p. 497
PROGESTERONE, TESTOSTERONE, NORGESTREL; NORETHINDRONE, ETHYNYL
ESTRADIOL; IMPLANT

516
SUSTAINED RELEASE OF SULPHAMETHIZOLE, 5-FLUOROURACIL AND
DOXORUBICIN FROM ETHYL CELLULOSE-POLYLACTIC ACID MICROCAPSULES
Itoh, M. et al.
Chem. Pharm. Bull., 28, 1980, p. 1051

517
LONG-TERM CONTROLLED DELIVERY OF LEVONORGESTREL IN RATS BY MEANS
OF SMALL BIODEGRADABLE CYLINDERS
Wise, D.L.
J. Pharm. Pharmacol., 32, 1980, p. 399

Lactic Acid — Glycolic Acid Copolymer

518
LACTIC/GLYCOLIC ACID POLYMERS AS NARCOTIC ANTAGONIST DELIVERY
SYSTEMS
Schwope, A.D. et al.
Life Sciences, 17, 1975, p. 1877
NALTREXONE

519
SUSTAINED RELEASE OF AN ANTIMALARIAL DRUG USING A COPOLYMER OF
GLYCOLIC LACTIC ACID
Wise, D.L. et al.
Life Sci., 19, 1976, p. 867
DIAMINONAPHTHYLSULPHONYL QUINAZOLINE

520
LACTIC/GLYCOLIC ACID POLYMERS
Wise, D.L. et al.
Drug Carriers in Biology and Medicine, 1979, p. 237
Gregoriadis, G. ed.

LACTIDE POLYMERS AND COPOLYMERS

Lactide Polymer

521
SUSTAINED RELEASE DEPOT FOR NARCOTIC ANTAGONISTS
Martin, W.R. & Sandquist, V.L.
Arch. Gen. Psychiatry., 30, 1974, p. 31
NALTREXONE MONOHYDRATE

522
HYDROLYTIC DEGRADATION OF POLY DL-(LACTIDE)
Mason, N.S. et al.
ACS Coatings & Plastics Preprints, 42, 1980, p. 436

Lactide-Glycolide Copolymer

523
LARGER ANIMAL TESTING OF AN INJECTABLE SUSTAINED RELEASE CONTROL SYSTEM
Gresser, J.D. et al.
Contraception, 17, 1978, p. 253
NORETHISTERONE

MALEIC ACID — BUTYL ETHYLENE COPOLYMER

524
SYNTHETIC POLYMERS AS POTENTIAL SUSTAINED-RELEASE COATINGS
Kleber, J.W. et al.
J. Pharm. Sci., 53, 1964, p. 1519
PREDNISOLONE

MALEIC ANHYDRIDE POLYMER

525
COMPARATIVE STUDY OF ANTITUMOUR AND TOXICOLOGIC PROPERTIES OF RELATED POLYANIONS
Ottenbrite, R. et al.
Polymer, 18, 1977, p. 461

526
CONTROLLED RELEASE BY POLYMER DISSOLUTION. I. PARTIAL ESTERS OF
MALEIC ANHYDRIDE COPOLYMERS, PROPERTIES AND THEORY
Heller, J. et al.
J. Appl. Polym. Sci., 22, 1978, p. 1991

MELAMINE FORMALDEHYDE RESIN

527
MICROENCAPSULATION OF PHENOBARBITAL BY SPRAY POLYCONDENSATION
Voellmy, C. et al.
J. Pharm. Sci., 66, 1977, p. 631

METHACRYLAMIDE POLYMERS

Ethyl Methacrylamide Polymer

528
NEW TYPES OF SYNTHETIC INFUSION SOLUTIONS. I. INVESTIGATION OF THE EFFECT OF SOLUTIONS OF SOME HYDROPHILIC POLYMERS ON BLOOD
Kopecek, J. et al.
J. Biomed. Mater. Res., 7, 1973, p. 179

Hydroxypropyl Methacrylamide Polymer

529
NEW TYPES OF SYNTHETIC INFUSION SOLUTIONS. I. INVESTIGATION OF THE EFFECT OF SOLUTIONS OF SOME HYDROPHILIC POLYMERS ON BLOOD
Kopecek, J. et al.
J. Biomed. Mater. Res., 7, 1973, p. 179

530
DEGRADATION OF SIDE CHAINS OF N-(2-HYDROXYPROPYL) METHACRYLAMIDE COPOLYMERS BY LYSOSOMAL ENZYMES
Duncan, R. et al.
Biochim. Biophys. Res. Comm., 94, 1980, p. 284

531
IMMUNOSUPPRESSIVE EFFECTS OF A SYNTHETIC POLYMER POLY N-(2-HYDROXYPROPYL) METHACRYLAMIDE (DUXON)
Paluska, E. et al.
Folia Biol. (Praha), 26, 1980, p. 304

Methacrylamide Polymer

532
MODEL REACTIONS FOR SYNTHESIS OF PHARMACOLOGICALLY ACTIVE POLYMERS BY WAY OF MONOMERIC AND POLYMERIC REACTIVE ESTERS
Batz, H.G. et al.
Angew. Chem. Int. Edn., 11, 1972, p. 1103

533
PHARMACOLOGICALLY ACTIVE POLYMERS. 6. SULPHADIAZINE AND BORON DERIVATIVES AS POTENTIAL CARRIERS FOR POLYMERS INTO CANCER TISSUE
Bartulin, J. et al.
Makromol. Chem., 175, 1974, p. 1007

534
SYNTHETIC POLYMERS IN CHEMOTHERAPY: GENERAL PROBLEMS
Kalal, J. et al.
Polymeric Drugs, 1978, p. 131, Academic Press
Donaruma, L.G. & Vogl, O. eds.

535
SYNTHESIS AND REACTIONS OF HYDROPHILIC FUNCTIONAL MICROSPHERES
FOR IMMUNOLOGICAL STUDIES
Rembaum, A. et al.
J. Macromol. Sci. Chem., A13, 1979, p. 603

METHACRYLATE POLYMERS AND COPOLYMERS

Amino Ethyl Methacrylate Polymer

536
ANTICOAGULANT ACTIVITY OF SOME SULPHATE-CONTAINING POLYMERS OF
THE METHACRYLATE TYPE
Sorm, M. et al.
J. Polym. Sci. Polym. Symp., No. 66, 1979, p. 349

Butylene Glycol Dimethacrylate Polymer

537
RELEASE OF INORGANIC FLUORIDE ION FROM RIGID POLYMER MATRICES
Halpern, B.D. et al.
ACS Coatings & Plastics Preprints, 36, 1976, p. 341
c.f. ACS Symp. Series, 33, 1976, p. 135

Butyl Methacrylate Polymer

538
FACTORS AFFECTING WATER VAPOUR TRANSMISSION THROUGH FREE
POLYMER FILMS
Swarbrick, J. et al.
J. Pharm. Sci., 61, 1972, p. 1645

Diethylene Glycol Dimethacrylate Polymer

539
CONTROLLED RELEASE OF BIOFUNCTIONAL SUBSTANCES BY RADIATION-
INDUCED POLYMERISATION. I. RELEASE OF POTASSIUM CHLORIDE BY
POLYMERISATION OF VARIOUS VINYL MONOMERS
Yoshida, M. et al.
Polymer, 19, 1978, p. 1375
CAPSULE

540
CONTROLLED RELEASE OF BIOFUNCTIONAL SUBSTANCES BY RADIATION-
INDUCED POLYMERISATION. II. RELEASE OF POTASSIUM CHLORIDE FROM
POROUS POLYDIETHYLENE GLYCOL DIMETHACRYLATE
Yoshida, M. et al.
Polymer, 19, 1978, p. 1379

Diethylene Glycol Dimethacrylate — Methyl Methacrylate Copolymer

541
DRUG ENTRAPMENT FOR CONTROLLED RELEASE IN RADIATION-POLYMERISED
BEADS
Yoshida, M. et al.
J. Pharm. Sci., 68, 1979, p. 628
POTASSIUM CHLORIDE; BEADS

Dimethyl Amino Ethyl Methacrylate Polymer

542
SYNTHESIS AND REACTIONS OF HYDROPHILIC FUNCTIONAL MICROSPHERES
FOR IMMUNOLOGICAL STUDIES
Rembaum, A. et al.
J. Macromol. Sci. Chem., A13, 1979, p. 603

Ethylene Glycol Dimethacrylate Polymer

543
CERVICAL HYDROGEL DILATOR: A NEW DELIVERY SYSTEM FOR PROSTAGLAN-
DINS
Akkapeddi, M.K. et al.
Adv. Expl. Med. & Biol., 47, 1974, p. 165
Controlled Release of Biologically Active Agents
Tanquary, A.C. & Lacey, R.E. eds.
PROSTAGLANDIN E2; PROSTAGLANDIN F2-ALPHA

544
APPLICATION OF LATEX MICROSPHERES IN THE ISOLATION OF PLASMA
MEMBRANES
Lim, R.W.
Biochim. Biophys. Acta., 394, 1975, p. 377
ANTIBODY LABELLING

545
NEW IMMUNOLATEX SPHERES: VISUAL MARKERS OF ANTIGENS ON
LYMPHOCYTES FOR SCANNING ELECTRON MICROSCOPY
Molday, R.S. et al.
J. Cell. Biol., 64, 1975, p. 75

546
RELEASE OF INORGANIC FLUORIDE ION FROM RIGID POLYMER MATRICES
Halpern, B.D. et al.
ACS Coatings & Plastics Preprints, 36, 1976, p. 341
c.f. ACS Symp. Series, 33, 1976, p. 135

547
CONTROLLED RELEASE OF BIOFUNCTIONAL SUBSTANCES BY RADIATION-
INDUCED POLYMERISATION. I. RELEASE OF POTASSIUM CHLORIDE BY
POLYMERISATION OF VARIOUS VINYL MONOMERS
Yoshida, M. et al.
Polymer, 19, 1978, p. 1375
CAPSULE

548
APPLICATION OF MAGNETIC MICROSPHERES IN LABELLING AND SEPARATION
OF CELLS
Molday, R.S. et al.
Nature, 268, Aug. 4, 1977, p. 437

549
CONTROLLED RELEASE OF BIOFUNCTIONAL SUBSTANCES BY RADIATION-
INDUCED POLYMERISATION. II. RELEASE OF POTASSIUM CHLORIDE FROM
POROUS POLYDIETHYLENE GLYCOL DIMETHACRYLATE
Yoshida, M. et al.
Polymer, 19, 1978, p. 1379

550
PROGESTIN PERMEATION THROUGH POLYMER MEMBRANES. II. DIFFUSION
STUDIES ON HYDROGEL MEMBRANES
Zentner, G.M. et al.
J. Pharm. Sci., 67, 1978, p. 1353
PROGESTERONE

551
SYNTHESIS AND REACTIONS OF HYDROPHILIC FUNCTIONAL MICROSPHERES
FOR IMMUNOLOGICAL STUDIES
Rembaum, A. et al.
J. Macromol. Sci. Chem., A13, 1979, p. 603

Ethylene Glycol Methacrylate Polymer

552
CONTROLLED DRUG RELEASE FROM POLYMERIC DELIVERY DEVICES. IV: IN
VITRO-IN VIVO CORRELATION OF SUBCUTANEOUS RELEASE OF NORGESTOMET
FROM HYDROPHILIC IMPLANTS
Chien, Y.W. & Lau, E.P.K.
J. Pharm. Sci., 65, 1976, p. 488

Glyceryl Methacrylate Polymer

553
PERMEATION OF WATER THROUGH SOME HYDROGELS
Refojo, M.F.
J. Appl. Polym. Sci., 9, 1965, p. 3417
CORNEAL IMPLANT

Glycidyl Methacrylate Polymer

554
RECENT PROBLEMS CONCERNING FUNCTIONAL MONOMERS AND POLYMERS
CONTAINING NUCLEIC ACID BASES
Takemoto, K.
Polymeric Drugs, 1978, p. 103, Academic Press
Donaruma, L.G. & Vogl, O. eds.

555
ANTIMICROBIAL AND ANTIPARASITIC POLYMERS
Brierly, J.A. et al.
ACS Coatings & Plastics Preprints, 42, 1980, p. 432

Glycidyl Methacrylate — Methyl Methacrylate Copolymer

556
DRUG ENTRAPMENT FOR CONTROLLED RELEASE IN RADIATION-POLYMERISED
BEADS
Yoshida, M. et al.
J. Pharm. Sci., 68, 1979, p. 628
POTASSIUM CHLORIDE

Hydroxyethyl Methacrylate Polymer (PolyHEMA, Hydron)

557
PERMEATION OF WATER THROUGH SOME HYDROGELS
Refojo, M.F.
J. Appl. Polym. Sci., 9, 1965, p. 3417
CORNEAL IMPLANT

558
USE OF HYDROPHILIC CONTACT LENSES TO INCREASE OCULAR PENETRATION
OF TOPICAL DRUGS
Waltman, S.R. & Kaufman, H.E.
Invest. Ophthalmol., 9, 1970, p. 250
FLUORESCEIN

559
HYDRON: AN IMPLANTABLE DEVICE FOR THE DELIVERY OF ANTITUMOUR
CHEMOTHERAPY
Arlen, M. et al.
Arch. Surg., 105, 1972, p. 100
FLUOROURACIL

560
STRUCTURE AND PERMEABILITY OF POROUS FILMS OF POLYHYDROXYETHYL
METHACRYLATE
Haldon, R.A. & Lee, B.E.
Brit. Polym. J., 4, 1972, p. 491

561
PILOCARPINE THERAPY WITH SOFT CONTACT LENSES
Podos, S.M. et al.
Amer. J. Ophthalmol., 73, 1972, p. 336

562
INTERACTION OF UREA WITH POLYHYDROXYETHYL METHACRYLATE
HYDROGELS
Ratner, B.D. & Miller, I.F.
J. Polym. Sci., A1, 10, 1972, p. 2425

563
ASPECTS OF THREE TYPES OF HYDROGELS FOR BIOMEDICAL APPLICATIONS
Bruck, S.D.
J. Biomed. Mat. Res., 7, 1973, p. 353

564
TRANSPORT THROUGH CROSSLINKED POLYHYDROXYETHYL METHACRYLATE
HYDROGEL MEMBRANES
Ratner, B.D. & Miller, I.F.
J. Biomed. Mat. Res., 7, 1973, p. 353
UREA; SODIUM CHLORIDE; GLYCINE; GLUCOSE

565
CONSTRUCTION AND PROPERTIES OF HYDROGEL-GRAFT-COATED COPPER-
BEARING INTRAUTERINE DEVICES FOR RABBITS
Scott, H. et al.
Biomat. Med. Dev. Artif. Org., 1, 1973, p. 681

566
DIFFUSION OF ANTI-TUMOUR DRUGS THROUGH MEMBRANES FROM
HYDROPHILIC METHACRYLATE GELS
Drobnik, J. et al.
J. Biomed. Mat. Res., 8, 1974, p. 45
CYTEMBENA; 5-FLUOROURACIL; FTORAFUR; BUTOCIN; SODIUM CHLORIDE

567
GEL ENTRAPMENT OF ENZYMES: KINETIC STUDIES OF IMMOBILISED GLUCOSE
OXIDASE
Hinberg, I. et al.
Biotech. Bioeng., 16, 1974, p. 159

568
LATEX SPHERES AS MARKERS FOR STUDIES OF CELL SURFACE RECEPTORS BY
SCANNING ELECTRON MICROSCOPY
Molday, R.S. et al.
Nature, 249, May 3, 1974, p. 81

569
BIOCOMPATIBLE IMPLANTS FOR THE SUSTAINED ZERO-ORDER RELEASE OF
NARCOTIC ANTAGONISTS
Abrahams, R.A. & Ronel, S.H.
J. Biomed. Mater. Res., 9, 1975, p. 355
CYCLAZOCINE

570
SUSTAINED RELEASE OF HORMONES AND STEROIDS FROM POLY-2-
HYDROXYETHYL METHACRYLATE MEMBRANES
Andersson, J.M. et al.
ACS Polymer Preprints, 16, 1975, p. 376
HYDROCORTISONE

571
APPLICATION OF LATEX MICROSPHERES IN THE ISOLATION OF PLASMA MEMBRANES
Lim, R.W.
Biochim. Biophys. Acta, 394, 1975, p. 377
ANTIBODY LABELLING

572
NEW IMMUNOLATEX SPHERES: VISUAL MARKERS OF ANTIGENS ON LYMPHOCYTES FOR SCANNING ELECTRON MICROSCOPY
Molday, R.S. et al.
J. Cell. Biol., 64, 1975, p. 75

573
CONTINUOUS MONITORING OF BLOOD GLUCOSE USING GEL ENTRAPPED GLUCOSE OXIDASE
O'Driscoll, K.F. et al.
ACS Polym. Preprints, 16, 1975, p. 372

574
CONTROLLED RELEASE OF FLUORIDE FROM HYDROGELS FOR DENTAL APPLICATIONS
Tarwater, O.R. et al.
ACS Polym. Preprints, 16, 1975, p. 382

575
DIFFUSION OF WATER SOLUBLE DRUGS FROM INITIALLY DRY HYDROGELS
Good, W.R.
Midland Macromol. Inst., Polymeric Delivery Systems, 5th Int. Symp., Aug. 1976, p. 139
TRIPELENNAMINE

576
RELEASE OF INORGANIC FLUORIDE ION FROM RIGID POLYMER MATRICES
Halpern, B.D. et al.
ACS Coatings & Plastics Preprints, 36, 1976, p. 331
c.f. ACS Symp. Series, 33, 1976, p. 135

577
SUSTAINED RELEASE OF MACROMOLECULES FROM POLYMERS
Langer, R. & Folkman, J.
Midland Macromol. Inst., Polymeric Delivery Systems, 5th Int. Symp., Aug. 1976, p. 175
SOYBEAN TRYPSIN INHIBITOR; LYSOZYME; ALKALINE PHOSPHATASE; CATALASE; INSULIN; HEPARIN

578
POLYMERS FOR THE SUSTAINED RELEASE OF PROTEINS AND OTHER
MACROMOLECULES
Langer, R. & Folkman, J.
Nature, 263, Oct. 28, 1976, p. 797

579
LIGHT MICROSCOPE IDENTIFICATION OF MURINE B AND T CELLS BY MEANS OF
FUNCTIONAL POLYMERIC MICROSPHERES
Gordon, I.L. et al.
Cell. Immunol., 28, 1977, p. 307

580
USE OF POLYMERS FOR THE SUSTAINED RELEASE OF PROTEINS AND OTHER
LARGE MOLECULES
Langer, R. & Folkman, J.
ACS Polymer Preprints, 18, No. 2, 1977, p. 379
LYSOZYME; ALKALINE PHOSPHATASE

581
PILOCARPINE ADMINISTRATION BY CONTACT LENS
Marmion, V.J. & Yurdakul, S.
Trans. Ophthal. Soc. U.K., 97, 1977, p. 162

582
APPLICATION OF MAGNETIC MICROSPHERES IN LABELLING AND SEPARATION
OF CELLS
Molday, R.S. et al.
Nature, 268, Aug. 4, 1977, p. 437

583
EFFECT OF METHOTREXATE SORBED ON MODIFIED 2-
HYDROXYETHYLMETHACRYLATE CARRIERS IN MICE OF C3H STRAIN WITH A
SOLID GARDNER LYMPHOSARCOMA
Motycka, K. et al.
Neoplasma, 24, 1977, p. 271

584
TETRACYCLINE: RELEASE FROM HYDROPHILIC GEL CONTACT LENS AND
INTRAOCULAR PENETRATION
Praus, R. & Krejci, L.
Ophthal. Res., 9, 1977, p. 213

585
HUMAN B LYMPHOCYTES IN GIEMSA STAINED PREPARATIONS
Taylor, C.R. et al.
J. Immunol. Meth., 17, 1977, p. 81

586
POLYMERIC AFFINITY DRUGS FOR CARDIOVASCULAR, CANCER AND UROLITHIASIS THERAPY
Goldberg, E.P.
Polymeric Drugs, 1978, p. 239, Academic Press
Donaruma, L.G. & Vogl, O. eds.

587
IMMOBILISATION OF STREPTOMYCES PHAEOCHROMOGENES CELLS AT A HIGH CONCENTRATION BY RADIATION INDUCED POLYMERISATION OF GLASS FORMING MONOMER
Kumakura, M. et al.
Eur. J. Appl. Microbiol. Biotechnol., 6, 1978, p. 13

588
TREATMENT OF SOLID GARDNER LYMPHOSARCOMA WITH METHOTREXATE SORBED ON 2-HYDROXYETHYLMETHACRYLATE POLYMER IN COMBINATION WITH LEUKOVORIN
Motycka, K. et al.
Neoplasma, 25, 1978, p. 217

589
DISTRIBUTION AND PHARMACOKINETICS OF METHOTREXATE IN LOCALISED CHEMOTHERAPY OF SOLID GARDNER'S LYMPHOSARCOMA
Slavikova, V. et al.
Neoplasma, 25, 1978, p. 211

590
CONTROLLED RELEASE OF BIOFUNCTIONAL SUBSTANCES BY RADIATION-INDUCED POLYMERISATION. I. RELEASE OF POTASSIUM CHLORIDE BY POLYMERISATION OF VARIOUS VINYL MONOMERS
Yoshida, M. et al.
Polymer, 19, 1978, p. 1375

591
PROGESTIN PERMEATION THROUGH POLYMER MEMBRANES. I. DIFFUSION STUDIES ON PLASMA SOAKED MEMBRANES
Zentner, G.M. et al.
J. Pharm. Sci., 67, 1978, p. 1347
PROGESTERONE

592
PROGESTIN PERMEATION THROUGH POLYMER MEMBRANES. II. DIFFUSION STUDIES ON HYDROGEL MEMBRANES
Zentner, G.M. et al.
J. Pharm. Sci., 67, 1978, p. 1353
PROGESTERONE

593
ENZYME IMMOBILISATION BY RADIATION-INDUCED POLYMERISATION OF 2-HYDROXYETHYL METHACRYLATE AT LOW TEMPERATURES
Kaetsu, I. et al.
Biotechnol. Bioeng., 21, 1979, p. 847
ALPHA AMYLASE; GLUCOAMYLASE; CELLULASE; GLUCOSIDASE; GLUCOSE OXIDASE

594
LOCALISED COMBINED CHEMOTHERAPY OF SOLID GARDNER LYMPHOSARCOMA BY METHOTREXATE BOUND TO SYNTHETIC POLYMERIC CARRIERS
Motycka, K. et al.
J. Polym. Sci. Polym. Symp., No. 66, 1979, p. 195
DIBROMOAMINOPTERIN

595
GEL ENTRAPPED MULTISTEP SYSTEMS
O'Driscoll, K.F. & Mercer, D.G.
Contemporary Topics in Polymer Science, 3, 1979, p. 319
Shen, M. ed. Plenum
INVERTASE; GLUCOSE OXIDASE

596
MICROSCOPIC DETERMINATION OF THE PENETRATION OF PROTEINS AND POLYSACCHARIDES INTO POLYHYDROXYETHYL METHACRYLATE AND SIMILAR HYDROGELS
Refojo, M.F. & Leong, F.L.
J. Polym. Sci. Polym. Symp., No. 66, 1979, p. 227
LYSOZYME; BOVINE SERUM ALBUMIN; DEXTRAN

597
SYNTHESIS AND REACTIONS OF HYDROPHILIC FUNCTIONAL MICROSPHERES FOR IMMUNOLOGICAL STUDIES
Rembaum, A. et al.
J. Macromol. Sci. Chem. A13, 1979, p. 603

598
IMMOBILISATION OF ENZYMES BY RADIATION-INDUCED POLYMERISATION OF GLASS FORMING MONOMERS. I. IMMOBILISATION OF SOME ENZYMES BY POLYHYDROXYETHYL METHACRYLATE
Yoshida, M. et al.
Polymer, 20, 1979, p. 3
ALPHA-AMYLASE; GLUCOAMYLASE

599
IMMOBILISATION OF ENZYMES BY RADIATION-INDUCED POLYMERISATION OF
GLASS-FORMING MONOMERS. II. EFFECTS OF COOLING RATE AND SOLVENT
ON POROSITY AND ACTIVITY OF IMMOBILISED ENZYMES
Yoshida, M. et al.
Polymer, 20, 1979, p. 9
GLUCOAMYLASE; MALTOSE

600
CONTROLLED RELEASE OF 5-FLUOROURACIL FROM ADSORBENTS CONTAINING
MATRICES POLYMERISED BY RADIATION
Yoshida, M. et al.
Polymer, J., 11, 1979, p. 775

601
PROGESTIN PERMEATION THROUGH POLYMER MEMBRANES. III.
POLYMERISATION SOLVENT EFFECT OF PROGESTERONE PERMEATION
THROUGH HYDROGEL MEMBRANES
Zentner, G.M. et al.
J. Pharm. Sci., 68, 1979, p. 794

602
PROGESTIN PERMEATION THROUGH POLYMER MEMBRANES. IV. MECHANISM
OF STEROID PERMEATION AND FUNCTIONAL GROUP CONTRIBUTION TO
DIFFUSION THROUGH HYDROGEL FILMS
Zentner, G.M. et al.
J. Pharm. Sci., 68, 1979, p. 970

603
HYDROGELS IN BIOMEDICAL APPLICATIONS
Pedley, D.G. et al.
Brit. Polym. J., 12, 1980, p. 99

604
ON THE ANOMALOUS SORPTION BEHAVIOUR OF CHLORHEXIDINE WITH
POLY(2-HYDROXYETHYL METHACRYLATE)
Plant, B.S. et al.
J. Pharm. Pharmacol., 32, 1980, p. 525

605
IMMUNOMICROSPHERES: REAGENTS FOR CELL LABELLING AND SEPARATION
Rembaum, A. & Dreger, W.J.
Science, 208, Apr. 25, 1980, p. 364

606
INSULIN PERMEABILITY OF HYDROPHILIC POLYACRYLATE MEMBRANES
Sefton, M.V. & Nishimura, E.
J. Pharm. Sci., 69, 1980, p. 208

607
MECHANISM OF IMMOBILISATION OF ENZYMES BY RADIATION-INDUCED
POLYMERISATION OF GLASS-FORMING MONOMERS
Yoshida, M. et al.
J. Macromol. Sci. Chem., A14, 1980, p. 541

Hydroxyethyl Methacrylate-Acrylamide Copolymer
and
Hydroxyethyl Methacrylate-Diethylene Glycol Dimethacrylate Copolymer
and
Hydroxyethyl Methacrylate-Hexanediol Monomethacrylate Copolymer
and
Hydroxyethyl Methacrylate-Hydroxyethyl Acrylate Copolymer
and
Hydroxyethyl Methacrylate-Methyl Methacrylate Copolymer
and
Hydroxyethyl Methacrylate-Vinyl Pyrrolidone Copolymer

608
IMMOBILISATION OF ENZYMES BY RADIATION-INDUCED COPOLYMERISATION
OF 2-HYDROXYETHYL METHACRYLATE AND OTHER HYDROPHILIC OR
HYDROPHOBIC COMONOMERS
Yoshida, M. et al.
J. Macromol. Sci. Chem., A14, 1980, p. 555
GLUCOAMYLASE

Hydroxyethyl Methacrylate-Methyl Methacrylate Copolymer

609
CONTROLLED RELEASE OF TETRACYCLINE. I. IN VITRO STUDIES WITH A TRILAMINATE 2-HYDROXYETHYL METHACRYLATE-METHYL METHACRYLATE SYSTEM
Olanoff, L. et al.
J. Pharm. Sci., 68, 1979, p. 1147

610
DRUG ENTRAPMENT FOR CONTROLLED RELEASE IN RADIATION-POLYMERISED BEADS
Yoshida, M. et al.
J. Pharm. Sci., 68, 1979, p. 628
POTASSIUM CHLORIDE

Methoxyethoxyethyl Methacrylate Polymer

611
PROGESTIN PERMEATION THROUGH POLYMER MEMBRANES. II. DIFFUSION STUDIES ON HYDROGEL MEMBRANES
Zentner, G.M. et al.
J. Pharm. Sci., 67, 1978, p. 1353
PROGESTERONE

Methoxyethyl Methacrylate Polymer

612
PROGESTIN PERMEATION THROUGH POLYMER MEMBRANES. II. DIFFUSION STUDIES ON HYDROGEL MEMBRANES
Zentner, G.M. et al.
J. Pharm. Sci., 67, 1978, p. 1353
PROGESTERONE

Methyl Methacrylate Polymer (PMMA)

613
FURTHER STUDIES OF POLYMERS AS CARCINOGENIC AGENTS IN ANIMALS
Oppenheimer, B.S. et al.
Cancer Res., 15, 1955, p. 333

614
DISINTEGRATION OF PROTECTIVE-COATED TABLETS AS DETERMINED BY
URINARY EXCRETION IN HUMANS
Ida, T. et al.
J. Pharm. Sci., 52, 1963, p. 472
RIBOFLAVIN

615
TISSUE REACTION INDUCED IN GUINEA PIGS BY PARTICULATE POLYMETHYL
METHACRYLATE, POLYTHENE AND NYLON OF THE SAME SIZE RANGE
Stinson, N.E.
Brit. J. Exp. Pathol., 46, 1965, p. 135

616
BEAD POLYMERISATION TECHNIQUE FOR SUSTAINED-RELEASE DOSAGE FORMS
Khanna, S.C. et al.
J. Pharm. Sci., 59, 1970, p. 615
CHLORAMPHENICOL; CHLOROTHIAZIDE; SULPHANILAMIDE; PENTOBARBITAL;
HYDROCORTISONE; PAPAVERINE BASE; ETHYLPHENYLGLUTARIMIDE

617
IN VITRO RELEASE OF CHLORAMPHENICOL FROM POLYMER BEADS OF ALPHA-
METHACRYLIC ACID AND METHYL METHACRYLATE
Khanna, S.C. & Speiser, P.
J. Pharm. Sci., 59, 1970, p. 1398

618
INTERFERON-INDUCING POLYCARBOXYLATES: MECHANISM OF PROTECTION
AGAINST VACCINIA VIRUS INFECTION IN MICE
Billiau, A. et al.
Infect. Immun., 5, 1972, p. 854

619
APPLICATION OF LATEX MICROSPHERES IN THE ISOLATION OF PLASMA
MEMBRANES
Lim, R.W.
Biochim. Biophys. Acta, 394, 1975, p. 377
ANTIBODY LABELLING

620
NEW IMMUNOLATEX SPHERES: VISUAL MARKERS OF ANTIGENS ON
LYMPHOCYTES FOR SCANNING ELECTRON MICROSCOPY
Molday, R.S. et al.
J. Cell. Biol., 64, 1975, p. 75

621
RELEASE OF INORGANIC FLUORIDE ION FROM RIGID POLYMER MATRICES
Halpern, B.D. et al.
ACS Coatings & Plastics Preprints, 36, 1976, p. 341
c.f. ACS Symp. Series, 33, 1976, p. 135

622
IN VITRO STUDIES OF POLYMETHYL METHACRYLATE ADJUVANTS
Kreuter, J. & Speiser, P.P.
J. Pharm. Sci., 65, 1976, p. 1624
INFLUENZA VIRUS

623
APPLICATION OF MAGNETIC MICROSPHERES IN LABELLING AND SEPARATION
OF CELLS
Molday, R.S. et al.
Nature, 268, Aug. 4, 1977, p. 437

624
SYNTHESIS AND SOME PROPERTIES OF ANTITHROMBOGENIC POLYMERS
Plate, N.A.
Polymeric Drugs, 1978, p. 63, Academic Press
Donaruma, L.G. & Vogl, O. eds.

625
RELEASE OF GENTAMYCIN FROM POLYMETHYL METHACRYLATE BEADS. AN
EXPERIMENTAL AND PHARMOKINETIC STUDY
Wahlig, H. et al.
J. Bone Jt. Surg., Br. Vol., 60B, 1978, p. 270

626
CONTROLLED RELEASE OF BIOFUNCTIONAL SUBSTANCES BY RADIATION-
INDUCED POLYMERISATION. I. RELEASE OF POTASSIUM CHLORIDE BY
POLYMERISATION OF VARIOUS VINYL MONOMERS
Yoshida, M. et al.
Polymer, 19, 1978, p. 1375

627
CONTROLLED RELEASE OF DRUGS FROM HYDROGEL MATRICES
Hosaka, S. et al.
J. Appl. Polym. Sci., 23, 1979, p. 2089
ERYTHROMYCIN; ERYTHROMYCIN ESTOLATE

628
SYNTHESIS AND REACTIONS OF HYDROPHILIC FUNCTIONAL MICROSPHERES FOR IMMUNOLOGICAL STUDIES
Rembaum, A. et al.
J. Macromol. Sci. Chem., A13, 1979, p. 603

629
STUDIES IN ANTIBIOTIC BONE CEMENT
Wahlig, H. & Dingledein, E.
Plastics in Medicine and Surgery, June 1979, p. 14.1
Plastics and Rubber Institute
GENTAMICIN

630
BIOPHYSICAL EVALUATION OF THE TUMORIGENIC RESPONSE TO IMPLANTED POLYMERS
Habal, M.B. & Powell, R.D.
J. Biomed. Mater. Res., 14, 1980, p. 447

631
IMMUNOLOGICAL EVALUATION OF THE TUMORIGENIC RESPONSE TO IMPLANTED POLYMERS
Habal, M.B.
J. Biomed. Mater. Res., 14, 1980, p. 455

632
CONTROLLED RELEASE OF MULTI-COMPONENT CYTOTOXIC AGENTS FROM RADIATION POLYMERISED COMPOSITES
Kaetsu, I. et al.
Biomaterials, 1, 1980, p. 17
MITOMYCIN; ADRYAMYCIN; TETRAHYDROFURYL FLUOROURACIL

633
CONTROLLED SLOW RELEASE OF CHEMOTHERAPEUTIC DRUGS FOR CANCER FROM MATRICES PREPARED BY RADIATION POLYMERISATION AT LOW TEMPERATURES
Kaetsu, I.
J. Biomed. Mat. Res., 14, 1980, p. 185
MITOMYCIN; 5-FLUOROURACIL; BLEOMYCIN

634
IMMUNOMICROSPHERES: REAGENTS FOR CELL LABELLING AND SEPARATION
Rembaum, A. & Dreyer, W.J.
Science, 208, Apr. 25, 1980, p. 364

635
ANTICANCER DRUG CAPSULE WITH CONTROLLED RELEASE FOR A LONG TIME
AND CHANGE OF RELEASE RATE
Yamada, A. et al.
Trans. Am. Soc. Artif. Int. Org., 26, 1980, p. 514
MITOMYCIN-C; TETRAHYDROFURYL-5-FLUOROURACIL; ADRIAMYCIN

Methyl Methacrylate-Ethyl Acrylate-Vinyl Pyrrolidone Copolymer

636
CONTROLLED RELEASE OF DRUGS FROM HYDROGEL MATRICES
Hosaka, S. et al.
J. Appl. Polym. Sci., 23, 1979, p. 2089
ERYTHROMYCIN; ERYTHROMYCIN ESTOLATE

Methyl Phenyldioxaborinanyl Methyl Methacrylate Polymer

and

Pyrimidinyl Sulphamoyl Phenyl Methacrylate Polymer

637
PHARMACOLOGICALLY ACTIVE POLYMERS: SULPHADIAZINE AND BORON
DERIVATIVES AS POTENTIAL CARRIERS FOR POLYMERS INTO CANCER TISSUE
Bartulin, J. et al.
Makromol. Chem., 175, 1974, p. 1007

Sulphoxyethyl Methacrylate Polymer

638
ANTICOAGULENT ACTIVITY OF SOME SULPHATE-CONTAINING POLYMERS OF
THE METHACRYLATE TYPE
Sorm, M. et al.
J. Polym. Sci. Polym. Symp., No. 66, 1979, p. 349

Tetraethylene Glycol Dimethacrylate Polymer

639
PROGESTIN PERMEATION THROUGH POLYMER MEMBRANES. II. DIFFUSION
STUDIES ON HYDROGEL MEMBRANES
Zentner, G.M. et al.
J. Pharm. Sci., 67, 1978, p. 1353
PROGESTERONE

Triethylene Glycol Monomethacrylate Polymer

640
NEW TYPES OF SYNTHETIC INFUSION SOLUTIONS. I. INVESTIGATION OF THE
EFFECT OF SOLUTIONS OF SOME HYDROPHILIC POLYMERS ON BLOOD
Kopecek, J. et al.
J. Biomed. Mater. Res., 7, 1973, p. 179

Trimethylolpropane Trimethacrylate Polymer

641
CONTROLLED RELEASE OF BIOFUNCTIONAL SUBSTANCES BY RADIATION-
INDUCED POLYMERISATION. I. RELEASE OF POTASSIUM CHLORIDE BY
POLYMERISATION OF VARIOUS VINYL MONOMERS
Yoshida, M. et al.
Polymer, 19, 1978, p. 1375
CAPSULE

642
CONTROLLED RELEASE OF 5-FLUOROURACIL FROM ADSORBENTS CONTAINING
MATRICES POLYMERISED BY RADIATION
Yoshida, M. et al.
Polym. J., 11, 1979, p. 775

Trimethylolpropane Trimethacrylate-Methyl Methacrylate Copolymer

643
DRUG ENTRAPMENT FOR CONTROLLED RELEASE IN RADIATION-POLYMERISED BEADS
Yoshida, M. et al.
J. Pharm. Sci., 68, 1979, p. 628
POTASSIUM CHLORIDE

METHACRYLIC ACID POLYMERS AND COPOLYMERS

Methacrylic Acid Polymer

644
EFFECT OF INTERFERON, POLYACRYLIC AND POLYMETHACRYLIC ACIDS ON
TAIL LESIONS IN VACCINIA-INFECTED MICE
De Clercq, E. & De Somer, P.
Appl. Microbiol., 16, 1968, p. 1314

645
ANTIVIRAL ACTIVITY OF POLYACRYLIC AND POLYMETHACRYLIC ACIDS. I.
MODE OF ACTION IN VITRO
De Somer, P. et al.
J. Virol., 2, 1968, p. 878

646
ANTIVIRAL ACTIVITY OF POLYACRYLIC AND POLYMETHACRYLIC ACIDS. II.
MODE OF ACTION IN VIVO
De Somer, P. et al.
J. Virol., 2, 1968, p. 886

647
POLYMETHACRYLIC ACID: EFFECTS ON LYMPHOCYTE OUTPUT OF THE
THORACIC DUCT IN RATS
Ormai, S. & De Clercq, E.
Science, 163, 1969, p. 471

648
IN VITRO RELEASE OF CHLORAMPHENICOL FROM POLYMER BEADS OF ALPHA-
METHACRYLIC ACID AND METHYLMETHACRYLATE
Khanna, S.C. & Speiser, P.
J. Pharm. Sci., 59, 1970, p. 1398

649
LYMPHOCYTE-MOBILISING AGENTS: EFFECTS OF POLYMETHACRYLIC ACID ON
TRANSPLANTABLE LYMPHOMA IN RODENTS
Ormai, S. et al.
Eur. J. Cancer, 6, 1970, p. 365

650
INTERACTION OF DRUGS WITH POLYMERS. III. PHASE SEPARATION OF
POLYACIDS BY O-BENZOYLTHIAMINE DISULPHIDE HYDROCHLORIDE AND
GASTROINTESTINAL ABSORPTION OF O-BENZOYLTHIAMINE DISULPHIDE-
POLYACID COMPLEXES
Tanaka, N. et al.
Chem. Pharm. Bull., 18, 1970, p. 1083

651
HOT EXTRUDED DOSAGE FORMS. I. TECHNOLOGY AND DISSOLUTION KINETICS
OF POLYMERIC MATRICES
El.Egakey, M.A. et al.
Pharm. Acta Helv., 46, 1971, p. 31

652
POLYMERS CONTAINING PHENETHYLAMINES
Weiner, B.Z. et al.
J. Med. Chem., 15, 1972, p. 410

653
DOSAGE, PLASMA CONCENTRATION AND ANTIARRHYTHMIC EFFECT OF
PROCAINAMIDE IN SUSTAINED-RELEASE TABLETS
Arstila, M. et al.
Acta Med. Scand., 195, 1974, p. 217

654
APPLICATION OF LATEX MICROSPHERES IN THE ISOLATION OF PLASMA
MEMBRANES
Lim, R.W.
Biochim. Biophys. Acta., 394, 1975, p. 377
ANTIBODY LABELLING

655
NEW IMMUNOLATEX SPHERES: VISUAL MARKERS OF ANTIGENS ON
LYMPHOCYTES FOR SCANNING ELECTRON MICROSCOPY
Molday, R.S. et al.
J. Cell. Biol., 64, 1975, p. 75

656
APPLICATION OF MAGNETIC MICROSPHERES IN LABELLING AND SEPARATION
OF CELLS
Molday, R.S. et al.
Nature, 268, Aug. 4, 1977, p. 437

657
POTENTIAL STRUCTURE-ACTIVITY RELATIONSHIPS INDIGENOUS TO POLYMER
SYSTEMS
Donaruma, L.G. et al.
Polymeric Drugs, 1978, p. 349, Academic Press
Donaruma, L.G. & Vogl, O. eds.

658
ANTICOAGULENT ACTIVITY OF SOME SULPHATE-CONTAINING POLYMERS OF
THE METHACRYLATE TYPE
Sorm, M. et al.
J. Polym. Sci. Polym. Symp., No. 66, 1979, p. 349

659
SUSTAINED RELEASE TABLET FORMULATION OF PROPRANOLOL
HYDROCHLORIDE WITH EUDRAGIT
Jayaswal, S.B. et al.
Aust. J. Pharm. Sci., 9, 1980, p. 22

660
IMMUNOMICROSPHERES: REAGENTS FOR CELL LABELLING AND SEPARATION
Rembaum, A. & Dreyer, W.J.
Science, 208, Apr. 25, 1980, p. 364

661
INSULIN PERMEABILITY OF HYDROPHILIC POLYACRYLATE MEMBRANES
Sefton, M.V. & Nishimura, E.
J. Pharm. Sci., 69, 1980, p. 208

Methacrylic Acid-Ethylene Copolymer

662
ANTIMICROBIAL POLYMERS
Ackart, W.B. et al.
J. Biomed. Mat. Res., 9, 1975, p. 55

METHACRYLIC ESTER-DIMETHYLAMINOETHYL METHACRYLATE COPOLYMER

663
CONTROLLED DRUG RELEASE FROM SMALL PARTICLES ENCAPSULATED WITH
ACRYLIC RESINS
Lehmann, K. et al.
Midland Macromol. Inst., Polymeric Delivery Systems, 5th Int. Symp. Aug.
1976, p. 111
METHAQUALONE; PHENYLPROPANOLAMINE

METHACRYLOXYTROPONE POLYMER

664
POLY-2-METHACRYLOXYTROPONE. A SYNTHETIC BIOLOGICALLY ACTIVE POLYMER
Cornell, R.J. & Donaruma, L.G.
J. Polym. Sci., A3, 1965, p. 827

METHACRYLOYLOXYBENZOIC ACID POLYMER

665
ANTIINFLAMMATORY ACTIVITIES OF POLYMERS OF o-METHACRYLOYLOXY-BENZOIC ACID
Grimova, J. & Hrabak, F.
J. Biomed. Mater. Res., 12, 1978, p. 525

OXYETHYLENE POLYMERS AND COPOLYMERS

Oxyethylene Polymer

666
EVALUATION OF POLYMERIC MATERIALS. IV. GRANULATING AGENTS FOR COMPRESSED TABLETS
Willis, C.R. et al.
J. Pharm. Sci., 54, 1965, p. 366

667
POLYMER-DRUG INTERACTION: STABILITY OF AQUEOUS GELS CONTAINING NEOMYCIN SULPHATE
Heyd, A.
J. Pharm. Sci., 60, 1971, p. 1343

668
MODIFICATION OF FIBRIN BY SYNTHETIC POLYMERS
Kovacs, G. et al.
J. Polym. Sci. Polym. Symp., No. 66, 1979, p. 201

669
CROSSLINKED HYDROPHILIC GELS FROM ABA POLYOXYETHYLENE-POLYOXYPROPYLENE-POLYOXYETHYLENE BLOCK COPOLYMERIC SURFACTANTS
Al-Saden, A.A. et al.
Int. J. Pharm., 5, 1980, p. 317

670
INDUCTION OF LABOUR WITH A SUSTAINED-RELEASE PROSTAGLANDIN E2 VAGINAL PESSARY
Embrey, M.P. et al.
British Medical Journal, 281, 1980, p. 901

Oxyethylene-Oxypropylene Copolymer (Poloxamers)

671
AVAILABILITY OF DRUGS IN THE PRESENCE OF SURFACE-ACTIVE AGENTS. II. EFFECTS OF SOME OXYETHYLENE OXYPROPYLENE POLYMERS ON THE BIOLOGICAL ACTIVITY OF HEXETIDINE
Saski, W. & Shah, S.G.
J. Pharm. Sci., 54, 1965, p. 277

672
USE OF WATER SOLUBLE POLYMERS AS SILVER ION CARRIERS FOR FALLOPIAN
TUBE CLOSURE
Hsia, H.T. et al.
ACS Coatings & Plastics Preprints, 36, 1976, p. 350
c.f. ACS Symp. Series, 33, 1976, p. 147

Oxyethylene Ether Polymer

673
ANTITUBERCULOUS EFFECT OF CERTAIN SURFACE-ACTIVE POLYOXYETHYLENE
ETHERS IN MICE
Cornforth, J.W. et al.
Nature, 168, July 28, 1951, p. 150

674
ANTITUBERCULOUS EFFECTS OF CERTAIN SURFACE ACTIVE POLYOXYETHYLENE
ETHERS
Cornforth, J.W. et al.
Brit. J. Pharmac. Chemother., 10, 1955, p. 73

675
PREPARATION OF ANTITUBERCULOUS POLYOXYETHYLENE ETHERS OF
HOMOGENEOUS STRUCTURE
Cornforth, J.W. et al.
Tetrahedron, 29, 1973, p. 1659

Oxyethylene Stearate Polymer

676
MICELLAR SOLUBILISATION OF BARBITURATES. II. SOLUBILITIES OF CERTAIN
BARBITURATES IN POLYOXYETHYLENE STEARATES OF VARYING HYDROPHILIC
CHAIN LENGTH
Gouda, M.W. et al.
J. Pharm. Sci., 59, 1970, p. 1402
PHENOBARBITAL; BARBITAL; AMOBARBITAL; DIALLYLBARBITURIC ACID;
CYCLOBARBITAL

OXYPROPYLENE POLYMER

677
EFFECTS OF POLYOXYPROPYLENE 15 STEARYL ETHER AND PROPYLENE GLYCOL
ON PERCUTANEOUS PENETRATION RATE OF DIFLORASONE DIACETATE
Turi, J.S. et al.
J. Pharm. Sci., 68, 1979, p. 275

PHOSPHAZENE POLYMER

678
SYNTHESIS OF PLATINUM DERIVATIVES OF POLYMERIC AND CYCLIC
PHOSPHAZENES
Allcock, H.R. et al.
J. Amer. Chem. Soc., 99, 1977, p. 3984
CHEMOTHERAPY

679
POLYORGANOPHOSPHAZENES DESIGNED FOR BIOMEDICAL USE
Allcock, H.R.
ACS Organometallic Polymers Symp., 37, 1977, p. 529
PLATINUM; CANCER CHEMOTHERAPY

680
POLYPHOSPHAZENES AS CARRIER MOLECULES FOR CONTROLLED RELEASE
SYSTEMS
Allcock, H.R. et al.
ACS Polym. Preprints, 21, 1980, p. 111
DESOXOESTRONE; ESTRONE; ESTRADIOL METHYL ETHER

681
ORGANOMETALLIC POLYMERS AS CHEMOTHERAPEUTIC DRUG DELIVERY
AGENTS
Carraher, C.E.
ACS Coating & Plastics Preprints, 42, 1980, p. 427
CIS-DICHLORODIAMINE PLATINUM

PROPYLENE POLYMERS AND COPOLYMERS

Propylene Polymer (PP)

682
BIOLOGICAL ACTIVITY OF POLYMERS
Conning, D.M.
Plastics in Medicine & Surgery Symposium, 15 & 16 Sept. 1971

683
SURFACE BONDED HEPARIN
Falb, R.D.
Polym. Sci. Technol., 8, 1974, p. 77
Johnson and Johnson Symposium, N.J., July 1974

684
ANTIMICROBIAL POLYMERS
Ackart, W.B. et al.
J. Biomed. Mater. Res., 9, 1975, p. 55
BENZALKONIUM SALTS; 8-HYDROXYQUINOLINIUM

685
SYNTHESIS AND SOME PROPERTIES OF ANTITHROMBOGENIC POLYMERS
Plate, N.A.
Polymeric Drugs, 1978, p. 63, Academic Press
Donaruma, L.G. & Vogl, O. eds.

686
SPECIFICITY OF POLYMER DEGRADATION IN THE LIVING BODY
Moiseev, Yu.V. et al.
J. Polym. Sci. Polym. Symp., No. 66, 1979, p. 269

Propylene-Maleic Anhydride Copolymer

687
INTERFERON-STIMULATING AND IN VIVO ANTIVIRAL EFFECTS OF VARIOUS
SYNTHETIC ANIONIC POLYMERS
Merigan, T.C. & Finkelstein, M.S.
Virology, 35, 1968, p. 363

Propylene Glycol Polymer

688
POSSIBLE COMPLEX FORMATION BETWEEN MACROMOLECULES AND CERTAIN PHARMACEUTICALS. X.
Guttman, D. & Higuchi, T.
J. Am. Pharm. Assn., 45, 1956, p. 659
PHENOL; RESORCINOL; CATECHOL; HYDROQUINONE; TANNIC ACID; PYROGALLOL; BENZOIC ACID; HYDROXYBENZOIC ACID; SALICYLIC ACID; SODIUM SALICYLATE; METHYL PARABEN; BETA NAPHTHOL; PICRIC ACID

689
PERMEATION OF WATER THROUGH SOME HYDROGELS
Refojo, M.F.
J. Appl. Polym. Sci., 9, 1965, p. 3417
CORNEAL IMPLANT

690
MACROMOLECULAR STEROID DRUGS FROM TESTOSTERONE OR BETA-ESTRADIOL AND POLYPROPYLENE GLYCOL
Kronick, P.L. & Daley, E.
Pharmacol. Res. Comm., 9, 1977, p. 279

691
VEHICLE EFFECTS IN PERCUTANEOUS ABSORPTION: IN VITRO STUDY OF IN-FLUENCE OF SOLVENT POWER AND MICROSCOPIC VISCOSITY OF VEHICLE ON BENZOCAINE RELEASE FROM SUSPENSION HYDROGELS
DiColo, G. et al.
J. Pharm. Sci., 69, 1980, p. 387

SILICONE RUBBER

692
FURTHER STUDIES OF POLYMERS AS CARCINOGENIC AGENTS IN ANIMALS
Oppenheimer, B.S. et al.
Cancer Res., 15, 1955, p. 333

693
DRUG PACEMAKERS IN THE TREATMENT OF HEART BLOCK
Folkman, J. & Long, D.M.
Ann. N.Y. Acad. Sci., 111, 1964, p. 857
TRIIODOTHYRONINE; IMPLANT

694
USE OF SILICONE RUBBER AS A CARRIER FOR PROLONGED DRUG THERAPY
Folkman, J. & Long, D.M.
J. Surg. Res., 4, 1964, p. 139
ISOPRENALIN; DIGITOXIN

695
CANCER INDUCTION BY POLYURETHANE AND POLYSILICONE PLASTICS
Hueper, W.C.
J. Nat. Cancer Inst., 33, 1964, p. 1005

696
EVALUATION OF POLYMERIC MATERIALS. I. SCREENING OF SELECTED
POLYMERS AS FILM COATING AGENTS
Munden, B.J. et al.
J. Pharm. Sci., 53, 1964, p. 395

697
PROLONGED ADMINISTRATION OF ATROPINE OR HISTAMINE IN A SILICONE
RUBBER IMPLANT
Bass, P. et al.
Nature, 208, Nov. 6, 1965, p. 591

698
USE OF SILICONE RUBBER IMPLANTS FOR THE SUSTAINED RELEASE OF
ANTIMALARIAL AND ANTISCHISTOSOMAL AGENTS (ABSTRACT)
Powers, K.G.
J. Parasitol., 51, 1965, p. 53
PYRIMETHAMINE; ANTIMONY DIMERCAPTOSUCCINATE

699
PASSAGE OF STEROIDS THROUGH SILICONE RUBBER
Dziuk, P.J. & Cook, B.
Endocrinology, 78, 1966, p. 208
ESTRADIOL; PROGESTERONE; ANDROSTENEDIONE; TESTOSTERONE; CORTISOL;
METHYLACETOXYPROGESTERONE; MELENGESTROL ACETATE; IMPLANT

700
SILICONE RUBBER: A NEW DIFFUSION PROPERTY USEFUL FOR GENERAL
ANAESTHESIA
Folkman, J. et al.
Science, 154, 1966, p. 148
ETHER; HALOTHANE; NITROUS OXIDE; CYCLOPROPANE

701
SUSTAINED RELEASE HORMONAL PREPARATIONS. I. DIFFUSION OF VARIOUS
STEROIDS THROUGH POLYMER MEMBRANES
Kincl, F.A. et al.
Steroids, 11, 1968, p. 673
NORPROGESTERONE; PROGESTERONE; TESTOSTERONE; NORETHINDRONE;
ESTRADIOL; MESTRANOL; CORTICOSTERONE; CORTISOL

702
SILASTIC ENTRAPMENT OF GLUCOSE OXIDASE-PEROXIDASE AND
ACETYLCHOLINE ESTERASE
Pennington, S.N. et al.
J. Biomed. Mat. Res., 2, 1968, p. 443

703
SUSTAINED RELEASE HORMONAL PREPARATIONS. 2. FACTORS CONTROLLING
THE DIFFUSION OF STEROIDS THROUGH DIMETHYLPOLYSILOXANE MEMBRANES
Sundaram, K. & Kincl, F.A.
Steroids, 12, 1968, p. 517
MELENGESTROL ACETATE; NORGESTREL; NORPROGESTERONE; CORTISOL

704
SUSTAINED RELEASE HORMONAL PREPARATIONS. 3. BIOLOGICAL EF-
FECTIVENESS OF 6-METHYL-17-ALPHA-ACETOXYPREGNA-4,6-DIENE-3,20-DIONE
Chang, C.C. & Kincl, F.A.
Steroids, 12, 1968, p. 689
MEGESTROL ACETATE; IMPLANT

705
PRELIMINARY STUDIES ON THE EFFECT OF HORMONE RELEASING INTRAUTERINE DEVICES
Doyle, L.L. & Clewe, T.H.
Amer. J. Obstet. Gynecol., 101, 1968, p. 564
MELENGESTROL ACETATE

706
DIFFUSION OF ANAESTHETICS AND OTHER DRUGS THROUGH SILICONE RUBBER: THERAPEUTIC IMPLICATIONS
Folkman, J. & Mark, V.H.
Trans. N.Y. Acad. Sci., 30, 1968, p. 1187
ETHER; HALOTHANE; METHOXYFLURANE

707
EVALUATION, CONTROL, AND PREDICTION OF DRUG DIFFUSION THROUGH POLYMERIC MEMBRANES. I.
Garrett, E.R. & Chemburkar, P.B.
J. Pharm. Sci., 57, 1968, p. 944
AMINOPROPIOPHENONE

708
FERTILITY CONTROL IN WOMEN WITH A PROGESTOGEN RELEASED IN MICROQUANTITIES FROM SUBCUTANEOUS CAPSULES
Croxatto, H. et al.
Amer. J. Obstet. Gynecol., 105, 1969, p. 1135
MEGESTROL ACETATE; IMPLANT

709
CHRONIC ANALGESIA BY SILICONE RUBBER DIFFUSION
Folkman, J. et al.
Surgery, 66, 1969, p. 194
METHOXYFLURANE

710
ACCEPTABILITY OF LONG TERM CONTRACEPTIVE STEROID ADMINISTRATION BY SUBCUTANEOUS SILASTIC CAPSULE
Tatum, H.J. et al.
Amer. J. Obstet. Gynecol., 105, 1969, p. 1139
MEGESTROL ACETATE; IMPLANT

711
SUSTAINED RELEASE HORMONAL PREPARATIONS. 5. ABSORPTION OF 6-METHYL-17-ALPHA-ACETOXYPREGNA-4,6-DIENE-3,20-DIONE FROM POLYDIMETHYL SILOXANE IMPLANTS IN VIVO
Benagiano, G. et al.
Acta Endocrinol., 63, 1970, p. 29
MEGESTROL ACETATE

712
CHEMISTRY AND PROPERTIES OF THE MEDICAL GRADE SILICONES
Braley, S.
J. Macromol. Sci. Chem., A4, 1970, p. 529

713
EFFECTS OF SUBCUTANEOUS IMPLANTATION OR INTRAUTERINE INSERTION OF SILASTIC TUBE CONTAINING STEROIDS ON THE FERTILITY OF RATS
Casas, J.H. & Chang, M.C.
Biol. Reprod., 2, 1970, p. 315
CHLORMADINONE ACETATE; ESTRADIOL; ETHINYL ESTRADIOL; PROGESTERONE

714
SUSTAINED RELEASE HORMONAL PREPARATIONS. IV. BIOLOGIC EFFECTIVENESS OF STEROID HORMONES
Chang, C.C. & Kincl., F.A.
Fertil. & Steril., 21, 1970, p. 134
ESTRONE; PROGESTERONE; TESTOSTERONE; NORPROGESTERONE; MEGESTROL ACETATE

715
PREVENTION OF PREGNANCY IN THE RABBIT BY SUBCUTANEOUS IMPLANTATION OF SILASTIC TUBING CONTAINING OESTROGEN
Chang, M.C. et al.
Nature, 226, June 27, 1970, p. 1262

716
RELEASE, EXCRETION, TISSUE UPTAKE AND BIOLOGICAL EFFECTIVENESS OF ESTRADIOL FROM SILASTIC DEVICES IMPLANTED IN RATS
Cornette, J.C. & Duncan, G.W.
Contraception, 1, 1970, p. 339

717
LONG TERM CONTRACEPTION BY SUBCUTANEOUS SILASTIC CAPSULES CONTAINING MEGESTROL ACETATE
Coutinho, E.M. et al.
Contraception, 2, 1970, p. 313

718
EXCRETION OF MEGESTROL ACETATE RELEASED FROM SUBCUTANEOUS SILASTIC IMPLANTS IN WOMEN
Coutinho, E.M. et al.
J. Reprod. Fertil., 23, 1970, p. 345

719
PROGESTATIONAL ACTIVITY OF MEGESTROL ACETATE PDMS CAPSULES IN RHESUS MONKEYS
Cuadros, A. et al.
Contraception, 2, 1970, p. 29

720
MICRODOSE INTRAUTERINE PROGESTAGEN ASSOCIATED WITH INTRAUTERINE CONTRACEPTIVE DEVICES
Horne, H.W. et al.
Int. J. Fertil., 15, 1970, p. 210
MEGESTROL ACETATE

721
SUSTAINED RELEASE HORMONAL PREPARATIONS. 8
Kincl, F.A.
Acta Endocrinol., 64, 1970, p. 253
NORPROGESTERONE; IMPLANT

722
SUSTAINED RELEASE HORMONAL PREPARATIONS 9. PLASMA LEVELS AND AC-CUMULATION INTO VARIOUS TISSUES OF 6-METHYL-17-ALPHA-ACETOXY-4,6-PREGNADIENE-3,20-DIONE AFTER ORAL ADMINISTRATION OR ABSORPTION FROM POLYDIMETHYLSILOXANE IMPLANTS
Kincl, F.A. et al.
Acta Endocrinol., 64, 1970, p. 508
MEGESTROL ACETATE

723
SUSTAINED RELEASE HORMONAL PREPARATIONS. 6. PERMEABILITY CONSTANT OF VARIOUS STEROIDS
Kratochvil, P. et al.
Steroids, 15, 1970, p. 505
NORPROGESTERONE; PROGESTERONE; TESTOSTERONE; MEGESTROL ACETATE; MELENGESTROL ACETATE; NORGESTREL; NORETHINDRONE; ESTRADIOL; MESTRANOL; CORTISOL; IMPLANT

724
DIFFUSION OF PROGESTOGENS THROUGH SILASTIC RUBBER IMPLANTS
Lifchez, A.S. & Scomegna, A.
Fertil. & Steril., 21, 1970, p. 426
PROGESTERONE; NORGESTREL; CHLORMADINONE ACETATE

725
EXPERIMENTAL ERRORS RESULTING FROM UPTAKE OF LIPOPHILIC DRUGS BY
SOFT PLASTIC MATERIALS
Minder, R. et al.
Biochem. Pharmacol., 19, 1970, p. 2179
IMIPRAMINE

726
CONTRACEPTIVE EFFECT OF VARYING DOSES OF PROGESTOGEN IN SILASTIC
VAGINAL RINGS
Mishell, D.R. & Lumkin, M.E.
Fertil. & Steril., 21, 1970, p. 99
PROGESTERONE

727
CONTRACEPTION BY MEANS OF A SILASTIC VAGINAL RING IMPREGNATED
WITH MEDROXYPROGESTERONE ACETATE
Mishell, D.R. et al.
Amer. J. Obstet. Gynecol., 107, 1970, p. 100

728
EFFECT OF MOLECULAR INTERACTION ON PERMEATION OF ORGANIC
MOLECULES THROUGH DIMETHYLPOLYSILOXANE MEMBRANE
Nakano, M. & Patel, N.K.
J. Pharm. Sci., 59, 1970, p. 77
SALICYLIC ACID; METHOXYCAFFEINE; CAFFEINE

729
RELEASE, UPTAKE AND PERMEATION BEHAVIOUR OF SALICYLIC ACID IN
OINTMENT BASES
Nakano, M. & Patel, N.K.
J. Pharm. Sci., 59, 1970, p. 985

730
FERTILITY CONTROL IN MALE RATS BY CONTINUOUS RELEASE OF
MICROQUANTITIES OF CYPROTERONE ACETATE FROM SUBCUTANEOUS
SILASTIC CAPSULES
Prasad, M.R.N. et al.
Contraception, 2, 1970, p. 165

731
RELEASE OF MEDROXYPROGESTERONE ACETATE FROM A SILICONE POLYMER
Roseman, T.J. & Higuchi, W.I.
J. Pharm. Sci., 59, 1970, p. 353

732
LONG TERM APPLICATION OF STEROIDS ENCLOSED IN DIMETHYL-POLYSILOXANE (SILASTIC): IN VITRO AND IN VIVO EXPERIMENTS
Schuhmann, R. & Taubert, H.D.
Acta Biol. Med. Germ. Band., 24, 1970, Seite 897
IN VITRO: PROGESTERONE; MEDROXYPROGESTERONE ACETATE; PROGESTERONE CAPRONATE; PROGESTERONE CHLORMADINONE ACETATE; NORGESTREL; CYPROTERONE ACETATE; GESTONOR CAPRONATE; METHYLNOR-TESTOSTERONE; ETHINYL NORTESTOSTERONE; ETHINYLNORTESTOSTERONE ACETATE, LYNOESTRENOL
IN VIVO: PROGESTERONE; MEDROXYPROGESTERONE ACETATE; NORGESTREL; CHLORMADINONE ACETATE; ETHINYLNORTESTOSTERONE; ETHINYLNOR-TESTOSTERONE ACETATE; CYPROTERONE ACETATE; ESTRADIOL; METHYLNOR-TESTOSTERONE

733
INTRAUTERINE ADMINISTRATION OF PROGESTERONE BY A SLOW RELEASING DEVICE
Scommegna, A. et al.
Fertil. & Steril., 21, 1970, p. 201

734
FERTILITY CONTROL AND ACCEPTABILITY IN WOMEN OF CONTRACEPTIVE STEROIDS RELEASED IN MICROQUANTITIES FROM SUBCUTANEOUS SILASTIC CAPSULES
Tatum, H.J.
Contraception, 1, 1970, p. 253
MEGESTROL ACETATE

735
SUSTAINED RELEASE HORMONAL PREPARATIONS
Zbuzkova, V. & Kincl, F.A.
Endocrinol. Exp., 4, 1970, p. 215
NORGESTREL

736
SUSTAINED RELEASE HORMONAL PREPARATIONS 12. PLASMA LEVELS OF 6-METHYL-17-ALPHA-ACETOXY-4,6-PREGNADIENE-3,20-DIONE IN HAMSTERS
Zbuzkova, V. & Kincl, F.A.
Steroids, 16, 1970, p. 447
MEGESTROL ACETATE; IMPLANT

737
CHEMISTRY AND PROPERTIES OF THE MEDICAL GRADE SILICONES
Braley, S.
Biomedical Polymers, 1971
Remaum, A. & Shen, M. eds.

738
SUPPRESSION OF PREGNANCY IN THE RABBIT BY SUBCUTANEOUS
IMPLANTATION OF SILASTIC TUBES CONTAINING VARIOUS ESTROGENIC
COMPOUNDS
Chang, M.C. et al.
Fertil. & Steril., 22, 1971, p. 383
ETHINYL ESTRADIOL; DIETHYLSTILBESTROL; QUINESTROL

739
CONTRACEPTIVE ACTION OF MEGESTROL ACETATE IMPLANTS IN WOMEN
Croxatto, H.B. et al.
Contraception, 4, 1971, p. 155

740
MEMBRANE DIFFUSION. II. INFLUENCE OF PHYSICAL ADSORPTION ON
MOLECULAR FLUX THROUGH HETEROGENEOUS DIMETHYLPOLYSILOXANE
BARRIERS
Flynn, G.L. & Roseman, T.J.
J. Pharm. Sci., 60, 1971, p. 1788
AMINOACETOPHENONE; ETHYL AMINOBENZOATE

741
DIFFUSION OF PROGESTERONE THROUGH INTRAUTERINE SILASTIC RUBBER
IMPLANTS
Gibor, Y.
Fertil. & Steril., 22, 1971, p. 671

742
IMPLANTED RESERVOIR OF MORPHINE SOLUTION FOR RAPID INDUCTION OF
PHYSICAL DEPENDENCE IN RATS
Goode, P.G.
Brit. J. Pharmac., 41, 1971, p. 558

743
STEROID RELEASE FROM SILICONE ELASTOMER CONTAINING EXCESS DRUG IN
SUSPENSION
Haleblian, J. et al.
J. Pharm. Sci., 60, 1971, p. 541
CHLORMADINONE ACETATE

744
SUSTAINED RELEASE HORMONAL PREPARATIONS
Kincl, F.A. & Rudel, H.W.
Acta Endocrinol., 66, 1971, Suppl. 151, p. 5
MEGESTROL ACETATE; IMPLANT

745
EFFECTS OF INTERACTION WITH SURFACTANTS, ADSORBENTS AND OTHER SUB-
STANCES ON THE PERMEATION OF CHLORPROMAZINE THROUGH A DIMETHYL
POLYSILOXANE MEMBRANE
Nakano, M.
J. Pharm. Sci., 60, 1971, p. 571

746
IN VIVO CHEMO DIFFUSION OF L-DOPA
Siegel, P. & Atkinson, J.R.
J. Appl. Physiol, 30, 1971, p. 900
MEMBRANE

747
CONTINUOUS STEROID TREATMENT BY SUBDERMAL POLYSILOXANE IMPLANTS
Benagiano, G. & Ermini, M.
Acta Europ. Fertil., 3, 1972, p. 119

748
DESIGN AND INITIAL TESTING OF AN IMPLANTABLE INFUSION PUMP
Blackshear, P.J.
Surg. Gynecol. Obstet., 134, 1972, p. 51
HEPARIN

749
FURTHER STUDIES ON LONG-TERM CONTRACEPTION BY SUBCUTANEOUS
SILASTIC CAPSULES CONTAINING MEGESTROL ACETATE
Coutinho, E.M. et al.
Contraception, 5, 1972, p. 389

750
CONTROL OF HAMSTER FERTILITY WITH PROSTAGLANDIN F2 ALPHA IMPLANTS
Davis, B.K. & Chang, M.C.
Acta Endocrinol., 70, 1972, p. 97

751
MEMBRANE DIFFUSION. III. INFLUENCE OF SOLVENT COMPOSITION AND
PERMANENT SOLUBILITY ON MEMBRANE TRANSPORT
Flynn, G.L. & Smith, R.W.
J. Pharm. Sci., 61, 1972, p. 61
AMINOACETOPHENONE

752
INHIBITION OF OVULATION WITH CYCLIC USE OF PROGESTOGEN-
IMPREGNATED INTRAVAGINAL DEVICES
Mishell, D.R. et al.
Am. J. Obstet. Gynecol., 113, 1972, p. 927

753
CO-PERMEATION ENHANCEMENT OF DRUG TRANSMISSION RATES THROUGH
SILICONE RUBBER
Most, C.F.
J. Biomed. Mat. Res., 6, 1972, p. 3
BENZOCAINE; IMPLANT

754
RELEASE OF STEROIDS FROM A SILICONE POLYMER
Roseman, T.J.
J. Pharm. Sci., 61, 1972, p. 46
PROGESTERONE;   HYDROXYPROGESTERONE;   METHYL
HYDROXYPROGESTERONE; METHYL ACETOXYPROGESTERONE

755
CONTINUOUS CANCER CHEMOTHERAPY: NITROSOUREA DIFFUSION THROUGH
IMPLANTED SILICON RUBBER CAPSULES
Schmidt, V. et al.
Trans. Am. Soc. Artif. Intern. Organs, 18, 1972, p. 45

756
INTERSTITIAL CANCER CHEMOTHERAPY WITH SILASTIC IMPLANTS
Sehgal, L.R.
Fed. Proc., 31, April 1972, p. 575

757
STUDIES ON SUSTAINED CONTRACEPTIVE EFFECTS WITH SUBCUTANEOUS
POLYDIMETHYLSILOXANE IMPLANTS. I. DIFFUSION OF MEGESTROL ACETATE IN
HUMANS
Benagiano, G. et al.
Acta Endocrinol., 73, 1973, p. 335

758
STUDIES ON SUSTAINED CONTRACEPTIVE EFFECTS WITH SUBCUTANEOUS
POLYDIMETHYLSILOXANE IMPLANTS
Benagiano, G. et al.
Acta Endocrinol., 73, 1973, p. 347
MEGESTROL ACETATE

759
STUDIES ON SUSTAINED CONTRACEPTIVE EFFECTS WITH SUBCUTANEOUS POLYDIMETHYLSILOXANE IMPLANTS. 3. FACTORS AFFECTING STEROID DIFFUSION IN VIVO AND IN VITRO
Ermini, M. et al.
Acta Endocrinol., 73, 1973, p. 360

760
DRUG-INCORPORATED SILICONE DISCS AS SUSTAINED RELEASE CAPSULES
Fu, J.C. et al.
J. Biomed. Mat. Res., 7, 1973, p. 71
CHLOROQUINE DIPHOSPHATE

761
DIFFUSION OF PYRIMETHAMINE FROM SILICONE RUBBER AND FLEXIBLE EPOXY DRUG CAPSULES
Fu, J.C. et al.
J. Biomed. Mater. Res., 7, 1973, p. 193

762
SUSTAINED RELEASE OF A HYPOGLYCEMIC SULPHONYLUREA FROM SILICONE RUBBER CAPSULES. IN VITRO STUDIES
Fu, J.C. et al.
Biomat. Med. Dev. Artif. Org., 1, 1973, p. 647

763
BASIC STUDIES FOR PROLONGED PROGESTOGEN ADMINISTRATION BY VAGINAL DEVICES
Henzl, M.R. et al.
Am. J. Obstet. Gynecol., 117, 1973, p. 101

764
LOCAL PROGESTATIONAL EFFECT OF NORGESTREL IN AN INTERUTERINE SILASTIC CAPSULE
Jones, R. et al.
Contraception, 8, 1973, p. 439

765
EVALUATION OF PROGESTERONE-CONTAINING SILICONE VAGINAL DEVICES IN RHESUS MONKEYS
Kirton, K.T. et al.
Contraception, 8, 1973, p. 561

766
DEPOT CANCER CHEMOTHERAPY THROUGH POLYMER MEMBRANES
Long, D.M. et al.
Rev. Surg., 30, 1973, p. 229
CYCLOPHOSPHAMIDE; MERCHLOROETHAMINE; TRIETHYLENE MELAMINE;
5-FLUORODEOXYURIDINE

767
DRUG PERMEATION THROUGH MEMBRANES. I. EFFECT OF VARIOUS SUB-
STANCES ON AMOBARBITAL PERMEATION THROUGH POLYDIMETHYLSILOXANE
Lovering, E.G. & Black, D.B.
J. Pharm. Sci., 62, 1973, p. 602

768
EFFECTIVENESS OF ANTITUMOUR AGENTS ADMINISTERED SUBCUTANEOUSLY
TO L1210 LEUKEMIC MICE IN SILICONE RUBBER DEVICES
Neil, G.L. et al.
Chemotherapy, 18, 1973, p. 27

769
PRODUCTION OF HYPERTENSION WITH DESOXYCORTICOSTERONE ACETATE
IMPREGNATED SILICONE RUBBER IMPLANTS
Ormsbee, H.S. & Ryan, C.F.
J. Pharm. Sci., 62, 1973, p. 255

770
CONTROL OF FERTILITY IN MALE RATS BY SUBCUTANEOUSLY IMPLANTED
SILASTIC CAPSULES CONTAINING TESTOSTERONE
Reddy, P.R.K. & Prasad, M.R.N.
Contraception, 7, 1973, p. 105

771
DIFFUSION IN VITRO AND IN VIVO OF 1-(2-CHLOROETHYL)-3-(TRANS-4-
METHYLCYCLOHEXYL)-1-NITROSOUREA FROM SILICONE RUBBER CAPSULES, A
POTENTIALLY NEW MODE OF CHEMOTHERAPY ADMINISTRATION
Rosenblum, M.L.
Cancer Res., 33, 1973, p. 906

772
CONTROLLED RELEASE OF TESTOSTERONE USING SILICONE RUBBER
Shippy, R.L. et al.
J. Biomed. Mat. Res., 7, 1973, p. 95

773
EVALUATION OF DIMETHYLPOLYSILOXANE FLUIDS AS VEHICLES FOR VARIOUS
PHARMACEUTICALS. I. EFFECT ON STABILITY AND DISSOLUTION OF ASPIRIN
Asker, A.F. & Whitworth, C.W.
J. Pharm. Sci., 63, 1974, p. 1630

774
CONTROLLED DRUG RELEASE FROM POLYMERIC DEVICES. I. TECHNIQUE FOR
RAPID IN VITRO RELEASE STUDIES
Chien, Y.W. et al.
J. Pharm. Sci., 63, 1974, p. 365
ETHYNODIOL DIACETATE

775
CONTROLLED DRUG RELEASE FROM POLYMERIC DELIVERY DEVICES. II. DIF-
FERENTIATION BETWEEN PARTITION-CONTROLLED AND MATRIX-CONTROLLED
DRUG RELEASE MECHANISMS
Chien, Y.W. & Lambert, H.J.
J. Pharm. Sci., 63, 1974, p. 515
ETHYNODIOL DIACETATE

776
ONE YEAR CONTRACEPTION WITH NORGESTRIENONE SUBDERMAL SILASTIC
IMPLANTS
Coutinho, E.M. & DaSilva, A.R.
Fertil. & Steril., 25, 1974, p. 170

777
SILICONE BASED RELEASE SYSTEMS
Duncan, G.W. et al.
Polym. Sci. Technol., 8, 1974, p. 205
Johnson & Johnson Symposium, N.J., July, 1974
PROGESTERONE; IMPLANT

778
SURFACE BONDED HEPARIN
Falb, R.D.
Polym. Sci. Technol., 8, 1974, p. 77
Johnson and Johnson Symposium, N.J. July, 1974

779
NICOTINE: RELEASE FROM SILICONE RUBBER IMPLANTS IN VIVO
Gaginella, T.S. & Bass, P.
Res. Commun. Chem. Pathol. Pharmacol., 7, 1974, p. 213

780
NICOTINE BASE PERMEATION THROUGH SILICONE ELASTOMERS: COM-
PARISON OF DIMETHYLPOLYSILOXANE AND TRIFLUOROPROPYLMETHYL-
POLYSILOXANE SYSTEMS
Gaginella, T.S. et al.
J. Pharm. Sci., 63, 1974, p. 1849

781
INTRAVAGINAL INSERTION OF A DIMETHYLPOLYSILOXANE-
POLYVINYLPYRROLIDONE-PROSTAGLANDIN F2 ALPHA TUBE FOR MID-TERM
ABORTION IN RABBITS
Lau, I.F. et al.
Am. J. Obstet. Gynecol., 120, 1974, p. 837

782
MID-TERM ABORTION WITH SILASTIC-PVP IMPLANT CONTAINING
PROSTAGLANDIN F2 ALPHA IN RABBITS, RATS AND HAMSTERS
Lau, I.F. et al.
Fertil. & Steril., 25, 1974, p. 839

783
DRUG PERMEATION THROUGH MEMBRANES. III. EFFECT OF pH AND VARIOUS
SUBSTRATES ON PERMEATION OF PHENYLBUTAZONE THROUGH EVERTED RAT
INTESTINE AND POLYDIMETHYLSILOXANE
Lovering, E.G. & Black, D.B.
J. Pharm. Sci., 63, 1974, p. 671

784
DRUG PERMEATION THROUGH MEMBRANES. IV. EFFECT OF EXCIPIENTS AND
VARIOUS ADDITIVES ON PERMEATION OF CHLORDIAZEPOXIDE THROUGH
POLYDIMETHYLSILOXANE MEMBRANES
Lovering, E.G. et al.
J. Pharm. Sci., 63, 1974, p. 1225

785
DIFFUSION LAYER EFFECTS ON PERMEATION OF PHENYLBUTAZONE THROUGH
PDMS
Lovering, E.G. & Black, D.B.
J. Pharm. Sci., 63, 1974, p. 1399

786
DEVELOPMENT OF A DELIVERY SYSTEM FOR PROSTAGLANDINS
Nuwayser, E.S. & Williams, D.L.
Adv. Expl. Med. Biol., 47, 1974, p. 145
Controlled Release of Biologically Active Agents
Tanquary, A.C. & Lacey, R.E. eds.
PROSTAGLANDIN F2 ALPHA

787
INFLUENCE OF SOLUTE PROPERTIES ON RELEASE OF P-AMINOBENZOIC ACID
ESTERS FROM SILICONE RUBBER: THEORETICAL CONSIDERATIONS
Roseman, T.J. & Yalkowsky, S.H.
J. Pharm. Sci., 63, 1974, p. 1639

788
SILICONE RUBBER: A DRUG DELIVERY SYSTEM FOR CONTRACEPTIVE STEROIDS
Roseman, T.J.
Advances in Experimental Medicine & Biology, 47, 1974, p. 99
Controlled Release of Biologically Active Agents
Tanquary, A.C. & Lacey, R.E. eds.
MEDROXYPROGESTERONE ACETATE

789
PROSTAGLANDIN E2: MID-TERM ABORTION IN RATS USING SILASTIC-PVP TUBES
Saksena, S.K. et al.
Prostaglandins, 7, 1974, p. 507

790
PROSTAGLANDIN F2 ALPHA IMPLANT-INDUCED ABORTION: EFFECT OF
PROGESTIN AND LUTENISING HORMONE CONCENTRATION AND ITS REVERSAL
BY PROGESTERONE IN RABBITS, RATS AND HAMSTERS
Saksena, S.V. et al.
Fertil. & Steril., 25, 1974, p. 845

791
FERTILITY CONTROL BY INTRAUTERINE RELEASE OF PROGESTERONE
Scommegna, A. et al.
Obstet. Gynecol., 43, 1974, p. 769

792
RECENT CLINICAL RESULTS WITH MEDROXYPROGESTERONE ACETATE IUDs
Stryker, J.C.
Intrauterine Devices: Development, Evaluation and Program
Implementation, 1974, p. 211
Wheeler, R.G. et al eds.
Academic Press

793
QUANTITATIVE ANALYTICAL METHOD FOR DETERMINATION OF DRUGS DISPER-
SED IN POLYMERS USING DIFFERENTIAL SCANNING CALORIMETRY
Theeuwes, F. et al.
J. Pharm. Sci., 63, 1974, p. 427
PROGESTERONE; CHOLESTEROL

794
LONG TERM CONTRACEPTION BY STEROID RELEASING IMPLANTS
Bhatnagar, S. et al.
Contraception, 11, 1975, p. 505
NORETHINDRONE ACETATE

795
COMPOSITE MEMBRANE ESTRADIOL IMPLANT
Bloch, R. et al.
J. Pharm. Sci., 64, 1975, p. 832

796
CONTROLLED DRUG RELEASE FROM POLYMERIC DELIVERY DEVICES. 3. IN
VITRO-IN VIVO CORRELATION FOR INTRAVAGINAL RELEASE OF ETHYNODIOL
DIACETATE FROM SILICONE DEVICES IN RABBITS
Chien, Y.W. et al.
J. Pharm. Sci., 64, 1975, p. 1776

797
CONTRACEPTIVE EFFECTIVENESS OF SILASTIC IMPLANTS CONTAINING THE
PROGESTIN R2323
Coutinho, E.M. et al.
Contraception, 11, 1975, p. 625

798
CLINICAL ASSESSMENT OF SUBDERMAL IMPLANTS OF MEGESTROL ACETATE D-
NORGESTREL AND NORETHINDRONE AS A LONG TERM CONTRACEPTIVE IN
WOMEN
Croxatto, H.B. et al.
Contraception, 12, 1975, p. 615

799
HORMONE-RELEASING SILICONE RUBBER INTRAUTERINE CONTRACEPTIVE
DEVICE; EFFECT OF INCORPORATION OF VARIOUS COMPOUNDS ON
INTRAUTERINE DEVICES IN RATS
Doyle, L.L.
Amer. J. Obstet. Gynecol., 121, 1975, p. 405

800
SELECTION OF STEROIDS FOR INCORPORATION INTO SILASTIC INTRAUTERINE
DEVICES
Doyle, L.L. et al.
J. Steroid Biochem., 6, 1975, p. 885
ISOXSUPRINE; PROGESTERONE; MELENGESTROL ACETATE;
MEDROXYPROGESTERONE ACETATE; NORGESTREL; NORETHINDRONE

801
INDOMETHACINE RELEASE FROM SILICONE RUBBER: A SUGGESTED METHOD
FOR CONTINUOUS INHIBITION OF PROSTAGLANDIN SYNTHESIS
Gaginella, T.S.
Res. Com. Chem. Pathol. & Pharmacol., 11, 1975, p. 323

802
PROLONGED RELEASE OF TESTOSTERONE PROPIONATE FROM NEW SOLID
STEROID SATURATED SILASTIC IMPLANTS IN RATS
Gupta, G.N.
Endocrinology, 96, (Suppl.), 1975, p. 244

803
EFFECT OF PROGESTIN R2323 RELEASED FROM VAGINAL RINGS ON OVARIAN
FUNCTION
Johansson, E.D.B. et al.
Contraception, 12, 1975, p. 299

804
INDUCTION OF TEMPORARY STERILITY IN FEMALE HAMSTERS BY AN
INTRAPERITONEAL SILASTIC-PVP-PGF2-ALPHA TUBE
Lau, I.F. et al.
Prostaglandins, 9, 1975, p. 893

805
STUDY OF BLOOD COAGULATION IN WOMEN TREATED WITH MEGESTROL
ACETATE IMPLANTS
Lira, P. et al.
Contraception, 12, 1975, p. 639

806
THERMODYNAMIC METHOD OF PREDICTING THE TRANSPORT OF STEROIDS IN
POLYMER MATRICES
Michaels, A.S. et al.
Amer. Inst. Chem. Eng. J., 21, 1975, p. 1073
PREGNADIENE METHYLDIONE; PROGESTERONE; NORPROGESTERONE;
TESTOSTERONE; PREGNATRIENE METHYL ETHINYLACETATE; ESTRADIOL;
NORGESTREL; NORETHINDRONE; CORTISOL; PREDNISOLONE; ESTRIOL

807
INITIAL CLINICAL STUDIES OF INTRAVAGINAL RINGS CONTAINING
NORETHINDRONE AND NORGESTREL
Mishell, D.R. et al.
Contraception, 12, 1975, p. 253

808
SUSTAINED RELEASE OF ANTIBIOTICS FROM SCLERAL BUCKLING MATERIALS. I.
GELATIN AND SOLID SILICONE RUBBER
Refojo, M.F. & Thomas, D.A.
Ophthalmic Res., 7, 1975, p. 33
CHLORAMPHENICOL; CHLORBUTANOL; LINCOMYCIN

809
CONTROLLED RELEASE FROM MATRIX SYSTEMS
Roseman, T.J.
J. Pharm. Sci., 64, 1975, p. 1731
ETHYNODIOL DIACETATE

810
DIFFUSION IN VITRO AND IN VIVO OF 1-(2-CHLOROETHYL)-3-(TRANS-4-
METHYLCYCLOHEXYL)-1-NITROSOUREA FROM SILICONE RUBBER CAPSULES: A
POTENTIAL NEW MODE OF CHEMOTHERAPY ADMINISTRATION
Rosenblum, M.L. et al.
Cancer Res., 33, 1975, p. 906

811
PLASMA CONCENTRATION OF A SYNTHETIC PROGESTIN R2323, RELEASED
FROM POLYSILASTIC VAGINAL RINGS
Viinikka, L. et al.
Contraception, 12, 1975, p. 309

812
INTRAVAGINAL CONTRACEPTION WITH A SYNTHETIC PROGESTIN R2323
Akinla, O. et al.
Contraception, 14, 1976, p. 671

813
ENZYMES IMMOBILISED BY MICROENCAPSULATION WITHIN SPHERICAL
ULTRATHIN POLYMERIC MEMBRANES
Chang, T.M.S.
J. Macromol. Sci. Chem., A10, 1976, p. 245
ERYTHROCYTE HAEMOLYSATE

814
THERMODYNAMICS OF CONTROLLED DRUG RELEASE FROM POLYMERIC
DELIVERY DEVICES
Chien, Y.W.
ACS Coatings & Plastics Preprints, 36, 1976, p. 326
c.f. ACS Symp. Series, 33, 1976, p. 53
NORGESTOMET

815
IN VITRO-IN VIVO CORRELATION ON CONTROLLED RELEASE OF DESOXYCOR-
TICOSTERONE ACETATE FROM MATRIX-TYPE SILICONE DEVICES
Chien, Y.W. et al.
ACS Coatings & Plastics Preprints, 36, 1976, p. 329
c.f. ACS Symp. Series, 33, 1976, p. 72
ETHYNODIOL DIACETATE

816
INTERFACING MATRIX RELEASE AND MEMBRANE ABSORPTION – ANALYSIS OF
STEROID ABSORPTION FROM A VAGINAL DEVICE IN THE RABBIT DOE
Flynn, G.L. et al.
ACS Coatings & Plastics Preprints, 36, 1976, p. 331
c.f. ACS Symp. Series, 33, 1976, p. 87
PROGESTERONE; ESTERONE; HYDROCORTISONE; TESTOSTERONE

817
UNIFIED MATHEMATICAL MODEL FOR DIFFUSION FROM DRUG-POLYMER COM-
POSITE TABLETS
Fu, J.C. et al.
J. Biomed. Mater. Res., 10, 1976, p. 743
PYRIMETHAMINE

818
CONTROLLED RELEASE OF DELMADINONE ACETATE FROM SILICONE POLYMER
TUBING: IN VITRO-IN VIVO CORRELATIONS
Kent, J.K.
ACS Coatings & Plastics Preprints, 36, 1976, p. 356
c.f. ACS Symp. Series, 33, 1976, p. 157

819
ABORTIFACIENT EFFECTS OF A NEWLY DEVELOPED VAGINAL SILASTIC DEVICE
Lauersen, N.H. & Wilson, K.H.
Contraception, 13, 1976, p. 697
METHYL PROSTAGLANDIN F2 ALPHA METHYL ESTER

820
DRUG PERMEATION THROUGH MEMBRANES. V. INTERACTION OF DIAZEPAM
WITH COMMON EXCIPIENTS
Lovering, E.G. et al.
J. Pharm. Sci., 65, 1976, p. 207

821
CONTROLLED DRUG PERMEATION. I. CONTROLLED RELEASE OF BUTAMBEN
THROUGH SILICONE MEMBRANE BY COMPLEXATION
Nakano, M. et al.
J. Pharm. Sci., 65, 1976, p. 709

822
D-NORGESTREL RELEASING IUD
Nilsson, C.G. et al.
Contraception, 13, 1976, p. 503

823
IMPORTANCE OF SOLUTE PARTITIONING ON THE KINETICS OF DRUG RELEASE
FROM MATRIX SYSTEMS
Roseman, T.J.
ACS Coatings & Plastics Preprints, 36, 1976, p. 321
MEDROXYPROGESTERONE ACETATE

824
MEDROXYPROGESTERONE ACETATE INTRAVAGINAL SILASTIC RING AS A
CONTRACEPTIVE DEVICE
Thiery, M. et al.
Contraception, 13, 1976, p. 605

825
INHIBITION OF THE POSITIVE FEEDBACK OF OESTRADIOL DURING TREATMENT
WITH SUBCUTANEOUS IMPLANTS OF D-NORGESTREL
Weiner, E. et al.
Contraception, 13, 1976, p. 287

826
CONTRACEPTION WITH MEGESTROL ACETATE IMPLANTS
Weiner, E. & Johansson, E.D.B.
Contraception, 13, 1976, p. 685

827
PLASMA LEVELS OF D-NORGESTREL, ESTRADIOL, AND PROGESTERONE DURING
TREATMENT WITH SILASTIC IMPLANTS CONTAINING D-NORGESTREL
Weiner, E. & Johansson, E.D.B.
Contraception, 14, 1976, p. 81

828
CONTRACEPTION WITH D-NORGESTREL SILASTIC RODS: PLASMA LEVEL OF
D-NORGESTREL AND INFLUENCE ON THE OVARIAN FUNCTION
Weiner, E. & Johansson, E.D.B.
Contraception, 14, 1976, p. 551

829
DEVELOPMENT OF AN ESTRIOL-RELEASING INTRAUTERINE DEVICE
Baker, R.W. et al.
ACS Coatings & Plastics Preprints, 37, 1977, p. 421

830
LONG-TERM CONTRACEPTION BY STEROID RELEASING IMPLANTS. II. A
PRELIMINARY REPORT ON LONG-TERM CONTRACEPTION BY SINGLE SILASTIC
IMPLANT RELEASING NORETHINDRONE ACETATE
Bhatnagar, S. et al.
Contraception, 15, 1977, p. 473

831
ZERO-ORDER RELEASE OF STEROIDS FROM VAGINAL DEVICES
Burton, F.G. et al.
ACS Coatings & Plastics Preprints, 37, 1977, p. 427
PROGESTERONE; NORETHINDRONE; NORGESTREL

832
ABORTION OF EARLY PREGNANCY ON AN OUTPATIENT BASIS USING SILASTIC
15(S)-15-METHYLPROSTAGLANDIN F2 ALPHA VAGINAL DEVICES
Corson, S.L. & Bolognese, R.J.
Fertil. & Steril., 28, 1977, p. 1056

833
SUSTAINED RELEASE OF ETHANOL FROM A SUBCUTANEOUS IMPLANT IN MICE
(ABSTRACT)
Erickson, C.K. et al.
Pharmacologist, 19, 1977, p. 167

834
CONVECTIVE DIFFUSIONAL ANALYSIS FOR DRUG TRANSPORT THROUGH A
TUBULAR POLYMERIC MEMBRANE
Flanagan, D.R. & Yalkowsky, S.H.
J. Pharm. Sci., 66, 1977, p. 337
BENZOCAINE; BUTAMBEN

835
POLYMERIC CONTRACEPTIVE DEVICES FOR CONTROLLED RELEASE OF
PROGESTINS
Gordon, N.R. et al.
Saf. Health Plast., Natl. Tech. Conf., Soc. Plast. Eng., 1977, p. 109
PROGESTIN; PROGESTERONE; NORETHINDRONE; NORGESTREL

836
LONG TERM CONTRACEPTION BY STEROID RELEASING IMPLANTS
Hillier, S.G. et al.
Contraception, 15, 1977, p. 473
NORETHINDRONE ACETATE

837
SLOW RELEASE SILICONE PELLET FOR CHRONIC AMPHETAMINE ADMINISTRATION
Huberman, H.S. et al.
Eur. J. Pharmacol., 45, 1977, p. 237

838
LOCAL TISSUE RESPONSE TO SILASTIC IMPLANT CONTAINING NORETHIN-DRONE ACETATE IN WOMEN
Jeyaseelan, S. et al.
Contraception, 15, 1977, p. 39

839
CONTROLLED DRUG PERMEATION. II. COMPARATIVE PERMEABILITY AND STABILITY OF BUTAMBEN AND BENZOCAINE
Juni, K. et al.
Chem. Pharm. Bull., 25, 1977, p. 1098

840
CONTROLLED RELEASE OF BUTAMBEN THROUGH SILICONE MEMBRANE BY MEANS OF COMPLEXATION AND MICELLAR SOLUBILISATION
Juni, K. et al.
Chem. Pharm. Bull., 25, 1977, p. 2807

841
CLINICAL PERFORMANCE AND ENDOCRINE PROFILES WITH CONTRACEPTIVE VAGINAL RINGS CONTAINING D-NORGESTREL
Mishell, D.R. et al.
Contraception, 16, 1977, p. 625

842
IODINATED SILICONE AS A CONTROLLED-RATE DELIVERY SYSTEM FOR MOLECULAR IODINE
Morian, W.D. & Vistnes, L.M.
Plast. Reconstr. Surg., 59, 1977, p. 216

843
EFFECT OF NORETHINDRONE ACETATE RELEASED FROM A SINGLE SILASTIC IM-PLANT ON SERUM FSH, LH, PROGESTERONE AND ESTRADIOL-17B OF WOMEN DURING THE FIRST 8 MONTHS OF TREATMENT
Rahman, S.A. et al.
Contraception, 16, 1977, p. 487

844
CONTRACEPTIVE RINGS: SELF-ADMINISTERED TREATMENT GOVERNED BY BLEEDING
Victor, A. & Johansson, E.D.B.
Contraception, 16, 1977, p. 137

845
IN VITRO AND IN VIVO CONSIDERATION OF A NOVEL MATRIX-CONTROLLED BOVINE PROGESTERONE-RELEASING INTRAVAGINAL DEVICE
Winkler, V.W. et al.
J. Pharm. Sci., 66, 1977, p. 816

846
RADIOIMMUNOASSAY OF OESTRONE IN PLASMA. PLASMA LEVELS OF OESTRONE AND OESTRADIOL IN OOPHORECTOMISED RHESUS MONKEYS DURING TREATMENT WITH SUBCUTANEOUS IMPLANTS CONTAINING OESTRONE
Axelsson, O. et al.
Acta Endocrinol., 87, 1978, p. 609

847
SUSTAINED RELEASE CHARACTERISTICS FOR ETHANOL AND OTHER VOLATILE SOLVENTS FROM SILASTIC TUBES IN MICE
Carlton, K. et al.
Fed. Proc., 37, 1978, p. 421

848
MICROSEALED DRUG DELIVERY SYSTEMS. I. IN VITRO – IN VIVO CORRELATION ON SUBCUTANEOUS RELEASE OF DESOXYCORTICOSTERONE ACETATE AND PROLONGED HYPERTENSIVE ANIMAL MODEL FOR CARDIOVASCULAR STUDIES
Chien, Y.W. et al.
J. Pharm. Sci., 67, 1978, p. 214

849
CLINICAL EXPERIENCE WITH IMPLANT CONTRACEPTION
Coutinkho, E.M.
Contraception, 18, 1978, p. 411

850
SALINE FILLED SILASTIC CAPSULES: A MORE SUSTAINED STEROID DELIVERY SYSTEM FOR INTRA-UTERINE CONTRACEPTION
Dileepkumar, S.A. et al.
J. Steroid Biochem., 9, 1978, p. 861
NORETHINDRONE ACETATE

851
SUSTAINED RELEASE OF ALCOHOL: SUBCUTANEOUS SILASTIC IMPLANTS IN MICE
Erickson, C.K. et al.
Science, 199, 1978, p. 1457
ETHANOL

852
RESPONSE OF LYMPHOID LEUKEMIA L1210 IN MICE TO IMPLANTABLE SUSTAINED RELEASE CYTOSINE ARABINOSIDE CAPSULES
Fu, J.C. et al.
J. Surg. Oncol., 10, 1978, p. 1

853
RESPONSE OF INTRAMUSCULAR WALKER 256 RAT TUMOUR TO SUSTAINED-RELEASE CYCLOPHOSPHAMIDE AND ARA-C CAPSULES
Fu, J.C.
J. Surg. Oncol., 10, 1978, p. 133
CYCLOPHOSPHAMIDE; CYTOSINE ARABINOSIDE; IMPLANT

854
CIRCULATING LEVELS OF NORETHINDRONE IN WOMEN WITH A SINGLE SILASTIC IMPLANT
Goyal, V. et al.
Contraception, 17, 1978, p. 375

855
RELEASE RATE OF TESTOSTERONE AND ESTROGENS FROM POLYDIMETHYL SILOXANE IMPLANTS FOR EXTENDED PERIODS IN VIVO COMPARED WITH LOSS IN VITRO
Greene, W.A. & Foote, R.H.
Int. J. Fertil., 23, 1978, p. 128
OESTRONE; ESTRADIOL; IMPLANT

856
CONTRACEPTION WITH LONG ACTING SUBDERMAL IMPLANTS. I. AN EF-FECTIVE AND ACCEPTABLE MODALITY IN INTERNATIONAL CLINICAL TRIALS
International Committee for Contraception Research of the Population Council
Contraception, 18, 1978, p. 315

857
LONG ACTING SILICONE DELIVERY SYSTEM FOR MORPHINE AND NALOXONE ADMINISTRATION
Isom, G.E. et al.
J. Pharm. Meth., 1, 1978, p. 121

858
EFFECT OF A SINGLE SILASTIC IMPLANT CONTAINING NORETHINDRONE
ACETATE ON HUMAN LACTATION
Jamwal, K. et al.
J. Steroid Biochem., 9, 1978, p. 860

859
RADIOIMMUNOASSAY OF NORETHINDRONE: SERUM LEVELS OF NORETHIN-
DRONE IN LACTATING WOMEN AFTER INSERTION OF A SINGLE SILASTIC
IMPLANT RELEASING NORETHINDRONE ACETATE
Laumas, V. et al.
Contraception, 18, 1978, p. 593

860
SUSTAINED RELEASE IMPLANTS OF CHEMICAL CARCINOGENS IN THE CANINE
TRACHEOBRONCHIAL TREE
Matsumura, K. et al.
Ann. Thorac. Surg., 25, 1978, p. 112
BENZOPYRENE; IMPLANT

861
CLINICAL PERFORMANCE AND ENDOCRINE PROFILES WITH CONTRACEPTIVE
VAGINAL RINGS CONTAINING A COMBINATION OF ESTRADIOL AND
D-NORGESTREL
Mishell, D.R. et al.
Am. J. Obstet. Gynecol., 130, 1978, p. 55

862
BLEEDING AND SERUM D-NORGESTREL, ESTRADIOL AND PROGESTERONE
PATTERNS IN WOMEN USING D-NORGESTREL SUBDERMAL POLYSILOXANE
CAPSULES FOR CONTRACEPTION
Moore, D.E. et al.
Contraception, 17, 1978, p. 315

863
STEROID RELEASE FROM SILASTIC CAPSULES AND RODS
Nash, H.A. et al.
Contraception, 18, 1978, p. 367
NORGESTREL; NORETHINDRONE; TESTOSTERONE; MEGESTROL ACETATE;
NORGESTRIENONE; R2323; NORETHANDROLONE

864
RELEASE OF CONTRACEPTIVE STEROIDS FROM SUSTAINED RELEASE DOSAGE
FORMS AND RESULTING PLASMA LEVELS
Nash, H.A. et al.
Contraception, 18, 1978, p. 395
LEVONORGESTREL; MEGESTROL ACETATE; NORGESTRIENONE

865
RELEASE OF A NITROSOUREA DERIVATIVE FROM REFILLABLE SILICONE RUBBER
IMPLANTS FOR THE TREATMENT OF INTRAOCULAR MALIGNANCIES
Refojo, M.F. et al.
J. Bioeng., 2, 1978, p. 437

866
SUSTAINED RELEASE ENDOBRONCHIAL CARCINOGENIC IMPLANTS
Shors, E.C. et al.
Surg. Forum., 29, 1978, p. 205
BENZOPYRENE

867
LONG TERM CONTRACEPTION BY A SINGLE SILASTIC IMPLANT – CONTAINING
NORETHINDRONE ACETATE IN WOMEN. A CLINICAL EVALUATION
Takkar, D. et al.
Contraception, 17, 1978, p. 341

868
PITUITARY AND GONADAL FUNCTION DURING THE USE OF NORGESTREL-
ESTRADIOL VAGINAL RINGS
Toivonen, J. et al.
Contraception, 18, 1978, p. 201

869
PROGESTIN PERMEATION THROUGH POLYMER MEMBRANES. I. DIFFUSION
STUDIES ON PLASMA-SOAKED MEMBRANES
Zentner, G.M. et al.
J. Pharm. Sci., 67, 1978, p. 1347
PROGESTERONE

870
ANTIBACTERIAL AND MECHANICAL PROPERTIES OF SILICONE RUBBER
CONTAINING GENTAMICIN SULPHATE
Bayston, R. & van Noort, R.
Plastics in Medicine and Surgery, June 1979, p. 25.1
Plastics and Rubber Institute
IMPLANT

871
SUSTAINED RELEASE HORMONAL PREPARATIONS FOR THE DELIVERY OF
FERTILITY-REGULATING AGENTS
Benagiano, G. et al.
J. Polym. Sci. Polym. Symp., No. 66, 1979, p. 129

872
LOW-LEVEL, PROGESTOGEN-RELEASING VAGINAL CONTRACEPTIVE DEVICES
Burton, F.G. et al.
Contraception, 19, 1979, p. 507
PROGESTERONE; NORETHINDRONE; NORGESTREL

873
LONG-TERM MAINTENANCE OF RECEPTIVITY BY SUBCUTANEOUS IMPLANTS OF ESTRADIOL
Campbell, C.S. & Baum, F.R.
Physiol. & Behav., 22, 1979, p. 1073

874
CONTROLLED DRUG RELEASE FROM POLYMERIC DELIVERY DEVICES. V. HYDROXY GROUP EFFECTS ON DRUG RELEASE KINETICS AND THERMODYNAMICS
Chien, Y.W. et al.
J. Pharm. Sci., 68, 1979, p. 689
PROGESTERONE; DESOXYCORTICOSTERONE; HYDROXYPROGESTERONE; CORTICOSTERONE; HYDROXYDESOXYCORTICOSTERONE; HYDROCORTISONE

875
EFFECT OF TEMPERATURE AND AN ION-EXCHANGE RESIN ON CATION DIFFUSION THROUGH SILICONE POLYMER TUBING
Christy, D.P. et al.
J. Pharm. Sci., 68, 1979, p. 1102

876
LONG TERM FERTILITY REGULATION WITH SUBDERMAL IMPLANTS
Diaz, S. & Croxatto, H.B.
Recent Advances in Reproduction and Regulation of Fertility, 1979, p. 559
Talwar, G.P. ed.
Elsevier/North-Holland Biomedical Press
LEVONORGESTREL

877
THREE-YEAR CLINICAL TRIAL WITH LEVONORGESTREL SILASTIC IMPLANTS
Diaz, S. et al.
Contraception, 19, 1979, p. 557

878
FIRST YEAR CLINICAL EXPERIENCE WITH SIX LEVONORGESTREL RODS AS SUBDERMAL CONTRACEPTION
Faundes, A. et al.
Contraception, 20, 1979, p. 167

879
PHARMACOKINETICS AND PHARMACODYNAMICS OF SUSTAINED RELEASE
SYSTEMS
Fotherby, K.
J. Steroid Biochem., 11, 1979, p. 457

880
INTRAVAGINAL AND INTRACERVICAL DEVICES FOR THE DELIVERY OF FERTILITY
REGULATING AGENTS
Gallegos, A.J.
J. Steroid Biochem., 11, 1979, p. 461
PROGESTERONE; NORETHISTERONE; LEVONORGESTREL; ESTRADIOL

881
PHARMACOKINETIC AND PHARMOACODYNAMIC EFFECTS OF SMALL DOSES OF
NORETHISTERONE RELEASED FROM VAGINAL RINGS CONTINUOUSLY DURING
90 DAYS
Landgren, B.M. et al.
Contraception, 19, 1979, p. 253

882
TERMINATION OF PSEUDOPREGNANCY IN RATS BY SILASTIC-PVP-PGF2 ALPHA
TUBE
Lau, I.F.
Prostaglandins Med., 2, 1979, p. 373

883
SUSTAINED RELEASE OF BCNU FOR THE TREATMENT OF INTRAOCULAR
MALIGNANCIES IN ANIMAL MODELS
Liu, H.S. et al.
Invest. Ophthalmol. Visual Sci., 18, 1979, p. 1061
NITROSOUREA

884
EFFECT OF WATER SOLUBLE CARRIERS ON MORPHINE SULPHATE RELEASE
FROM A SILICONE POLYMER
McGinty, J.W. et al.
J. Pharm. Sci., 68, 1979, p. 662

885
SPECIFICITY OF POLYMER DEGRADATION IN THE LIVING BODY
Moiseev, Yu.V. et al.
J. Polym. Sci. Polym. Symp., No. 66, 1979, p. 269

886
PLASMA LEVELS OF NORETHINDRONE AND EFFECT UPON OVARIAN FUNCTION DURING TREATMENT WITH SILASTIC IMPLANTS CONTAINING NORETHINDRONE
Odlind, V. et al.
Contraception, 19, 1979, p. 197

887
SUSTAINED RELEASE OF GENTAMICIN FROM PROSTHETIC HEART VALVES
Olanoff, L.S.
Trans. Am. Soc. Artif. Organs., 25, 1979, p. 334

888
SELECTIVITY OF SILICONE RUBBER TOWARD PROSTAGLANDIN PERMEABILITY
Roseman, T.J.
J. Pharm. Sci., 68, 1979, p. 263
PROSTAGLANDIN E2; PROSTAGLANDIN F2-ALPHA

889
METABOLIC AND ENDOCRINE STUDIES IN WOMEN USING NORETHINDRONE ACETATE IMPLANT
Shahani, S.M. et al.
Contraception, 19, 1979, p. 135

890
USE OF A CONTRACEPTIVE VAGINAL RING GOVERNED BY THE PATTERN OF INDIVIDUAL UTERINE BLEEDING
Toivonen, J. et al.
Contraception, 19, 1979, p. 401

891
EFFECTS OF SUSTAINED-RELEASE TESTOSTERONE ON MARKING BEHAVIOUR IN THE MONGOLIAN GERBIL
Turner, J.W.
Physiol. Behav., 23, 1979, p. 845

892
MAINTENANCE OF PHYSIOLOGIC CONCENTRATIONS OF PLASMA TESTOSTERONE IN THE CASTRATED MALE DOG, USING TESTOSTERONE-FILLED POLYDIMETHYLSILOXANE CAPSULES
Vincent, D.L. et al.
Am. J. Vet. Res., 40, 1979, p. 705

893
TREATMENT OF RHESUS MONKEYS WITH INTRAVAGINAL RINGS IMPREGNATED
WITH EITHER PROGESTERONE OR NORETHISTERONE
Wadsworth, P.F.
Contraception, 20, 1979, p. 339

894
TREATMENT OF RHESUS MONKEYS WITH INTRAVAGINAL RINGS LOADED WITH
LEVONORGESTREL
Wadsworth, P.F.
Contraception, 20, 1979, p. 559

895
ECTOPIC PREGNANCIES ASSOCIATED WITH LOW DOSE PROGESTAGEN-
RELEASING IUD's
Diaz, S. et al.
Contraception, 22, 1980, p. 259
MEGESTROL ACETATE; LEVONORGESTREL; NORETHINDRONE

896
VEHICLE EFFECTS IN PERCUTANEOUS ABSORPTION: IN VITRO STUDY OF IN-
FLUENCE OF SOLVENT POWER AND MICROSCOPIC VISCOSITY OF VEHICLE ON
BENZOCAINE RELEASE FROM SUSPENSION HYDROGELS
DiColo, G. et al.
J. Pharm. Sci., 69, 1980, p. 387

897
LACK OF EFFECT OF A PROGESTERONE-CONTAINING SILASTIC IUD ON
ENDOMETRIAL BLOOD FLOW IN RATS
Einer-Jensen, N.
Horm. Res., 12, 1980, p. 233

898
NORGESTREL-T IUD
El-Mahgoub, S.
Contraception, 22, 1980, p. 271

899
SUBCUTANEOUS SILASTIC IMPLANTS: MAINTENANCE OF HIGH BLOOD
ETHANOL LEVELS IN RATS DRINKING A LIQUID DIET
Erickson, C.K. et al.
Pharmacol. Biochem. Behav., 13, 1980, p. 781

900
FLUORIDE PROLONGED RELEASE PREPARATIONS FOR TOPICAL USE
Friedman, M.
J. Dent. Res., 59, 1980, p. 1392

901
BIOPHYSICAL EVALUATION OF THE TUMORIGENIC RESPONSE TO IMPLANTED POLYMERS
Habal, M.B. & Powell, R.D.
J. Biomed. Mater. Res., 14, 1980, p. 447

902
IMMUNOLOGICAL EVALUATION OF THE TUMORIGENIC RESPONSE TO IMPLANTED POLYMERS
Habal, M.B.
J. Biomed. Mater. Res., 14, 1980, p. 455

903
FOR INSULIN INFUSION: A MINIATURE PRECISION PERISTALTIC PUMP AND SILICONE RUBBER RESERVOIR
Jackman, W.S. et al.
Diabetes Care, 3, 1980, p. 322

904
CYPROTERONE ACETATE AND SEMINAL VESICLES IN THE REGULATION OF MALE FERTILITY
Nayyar, S.K. & Ghosh, T.
Contraception, 21, 1980, p. 183

905
SILICONE PELLET FOR LONG TERM CONTINUOUS ADMINISTRATION OF AMPHETAMINE
Nielson, E.B. & Ellison, G.
Commun. Psychopharmacol., 4, 1980, p. 17

906
PATTERNS OF OVULATION AND BLEEDING WITH A LOW LEVONORGESTREL-RELEASING INTRAUTERINE DEVICE
Nilsson, C.G. et al.
Contraception, 21, 1980, p. 155

907
LEVONORGESTREL PLASMA CONCENTRATION AND HORMONE PROFILES AFTER INSERTION AND AFTER ONE YEAR OF TREATMENT WITH A LEVONORGESTREL IUD
Nilsson, C.G. et al.
Contraception, 21, 1980, p. 225

908
SUSTAINED-RELEASE OF NIRIDAZOLE FROM SILICONE RUBBER IMPLANTS FOR
THE TREATMENT OF SCHISTOSOMA-MANSONI INFECTIONS
Olanoff, L.S. et al.
Amer. J. Trop. Med. & Hygiene, 29, 1980, p. 71

909
TEMPORAL RELATIONSHIP BETWEEN PROGESTIN AND LUTEINISING HORMONE
CONCENTRATIONS IN PSEUDOPREGNANT RATS TREATED WITH PROSTAGLAN-
DIN F2 ALPHA
Popkin, R. et al.
Prostaglandins Med., 4, 1980, p. 113

910
POLYDIMETHYLSILOXANE PELLETS FOR SUSTAINED DELIVERY OF MORPHINE IN
MICE
Riffee, W.H. et al.
J. Pharm. Sci., 69, 1980, p. 980

911
CLINICAL AND ENDOCRINOLOGICAL STUDY OF CONTINUOUS
LEVONORGESTREL ADMINISTRATION FROM SUBCUTANEOUS SOLID
POLYDIMETHYLSILOXANE RODS
Roy, S. et al.
Contraception, 21, 1980, p. 595

912
TESTOSTERONE REGULATION OF LUTEINISING HORMONE AND FOLLICLE
STIMULATING HORMONE SECRETION IN YOUNG MALE LAMBS
Schanbacher, B.D.
J. Anim. Sci., 51, 1980, p. 679

913
SUSTAINED RELEASE OF BENZOPYRENE FROM SILICONE POLYMER INTO THE
TRACHEOBRONCHIAL TREE OF HAMSTERS AND DOGS
Shors, E.C. et al.
Cancer Res., 40, 1980, p. 2288

914
SILASTIC IMPLANTS RELEASING OESTRONE IN THE TREATMENT OF
CLIMACTERIC COMPLAINTS
Sipinen, S. et al.
Maturitas, 2, 1980, p. 213

915
NORPLANT, REVERSIBLE IMPLANT CONTRACEPTION
Sivin, I. et al.
Stud. Fam. Plann., 11, 1980, p. 227
LEVONORGESTREL

916
SILICONES AND MEDICINE
Anon
Polymer News, 6, Jan. 1980, p. 116

STYRENE POLYMERS AND COPOLYMERS

Dimethylaminostyrene Polymer

917
EFFECTS OF A SERIES OF NEW SYNTHETIC HIGH POLYMERS ON CANCER METASTASES
Ferruti, P. et al.
J. Med. Chem., 16, 1973, p. 496

Styrene Polymer

918
FURTHER STUDIES OF POLYMERS AS CARCINOGENIC AGENTS IN ANIMALS
Oppenheimer, B.S. et al.
Cancer Res., 15, 1955, p. 333

919
EVALUATION OF POLYMERIC MATERIALS. I. SCREENING OF SELECTED POLYMERS AS FILM COATING AGENTS
Munden, B.J. et al.
J. Pharm. Sci., 53, 1964, p. 395

920
SUSTAINED RELEASE HORMONAL PREPARATIONS. I. DIFFUSION OF VARIOUS STEROIDS THROUGH POLYMER MEMBRANES
Kincl, F.A. et al.
Steroids, 11, 1968, p. 673
NORPROGESTERONE; PROGESTERONE; TESTOSTERONE; NORETHINDRONE; ESTRADIOL; MESTRANOL; CORTICOSTERONE; CORTISOL

921
MODEL REACTIONS FOR SYNTHESIS OF PHARMACOLOGICALLY ACTIVE POLYMERS BY WAY OF MONOMERIC AND POLYMERIC REACTIVE ESTERS
Batz, H.G. et al.
Angew. Chem. Int. Edn., 11, 1972, p. 1103

922
SUSPENSION POLYMERISATION FOR PREPARATION OF TIMED-RELEASE DOSAGE FORMS
Croswell, R.W. & Becker, C.H.
J. Pharm. Sci., 63, 1974, p. 443
ACETAMINOPHEN; BEAD

923
ANTIMICROBIAL POLYMERS
Ackart, W.B. et al.
J. Biomed. Mater. Res., 9, 1975, p. 55
BENZALKONIUM SALTS; 8-HYDROXYQUINOLINIUM

924
SLOW RELEASE OF WATER SOLUBLE SALTS FROM POLYMERS
Narkis, N. & Narkis, M.
J. Appl. Polym. Sci., 20, 1976, p. 3431
SODIUM CHLORIDE; SODIUM FLUORIDE

925
DRUG RELEASE FROM AND PROPERTIES OF POLYSTYRENE MICROCAPSULES
Nozawa, Y. & Higashide, F.
Midland Macromol. Inst., Polymeric Delivery Systems, 5th Int. Symp., Aug. 1976, p. 101
ALPHA-AMYLASE; SODIUM SALICYLATE

926
DRUG RELEASE FROM AND PROPERTIES OF POLYSTYRENE MICROCAPSULES
Nozawa, Y. et al.
J. Appl. Polym. Sci., 20, 1976, p. 3197
AMYLASE; SODIUM SALICYLATE

927
POLYMER-DRUG GRAFTS FOR IRON CHELATION
Ramirez, R.S. & Andrade, J.D.
J. Macromol. Sci. Chem., A10, 1976, p. 309
DEFEROXAMINE

928
USE OF Tc.99m-LABELLED TRIETHYLENETETRAMINE-POLYSTYRENE RESIN FOR MEASURING THE GASTRIC EMPTYING RATE IN HUMANS
Digenis, G.A. et al.
J. Pharm. Sci., 66, 1977, p. 442

929
POLYMER-BOUND CARBONIC ANHYDRIDES IN N-ACYLATION OF 7-AMINOCEPHALOSPORANIC ACID
Martin, G.E. et al.
J. Pharm. Sci., 67, 1978, p. 110

930
DISTRIBUTION OF RADIOLABELLED SUBVISIBLE MICROSPHERES AFTER INTRAVENOUS ADMINISTRATION TO BEAGLE DOGS
Schroeder, H.G. et al.
J. Pharm. Sci., 67, 1978, p. 504

931
PHYSIOLOGICAL EFFECTS OF SUBVISIBLE MICROSPHERES ADMINISTERED INTRAVENOUSLY TO BEAGLE DOGS
Schroeder, H.G. et al.
J. Pharm. Sci., 67, 1978, p. 509

932
RECENT PROBLEMS CONCERNING FUNCTIONAL MONOMERS AND POLYMERS CONTAINING NUCLEIC ACID BASES
Takemoto, K.
Polymeric Drugs, 1978, p. 103, Academic Press
Donaruma, L.G. & Vogl, O. eds.

933
SUSTAINED RELEASE OF CHLORPHENAMINE FROM MACROPOROUS KY-23 ION EXCHANGE RESIN BEADS
Geneidi, A.S. & Hamacher, H.
Pharm. Ind., 42, 1980, p. 198

934
IMMOBILISATION OF ANTIBODIES ON NYLON FOR USE IN ENZYME-LINKED IMMUNOASSAY
Hendry, R.M. & Herrman, J.E.
J. Immunol. Methods, 35, 1980, p. 285
IMMUNOGLOBULIN G; IMMUNOGLOBULIN E

935
CONTROLLED SLOW RELEASE OF CHEMOTHERAPEUTIC DRUGS FOR CANCER FROM MATRICES PREPARED BY RADIATION POLYMERISATION AT LOW TEMPERATURES
Kaetsu, I.
J. Biomed. Mat. Res., 14, 1980, p. 185
MITOMYCIN; 5-FLUOROURACIL; BLEOMYCIN

936
CLEARANCE OF Ce 141-LABELLED MICROSPHERES FROM BLOOD AND DISTRIBUTION IN SPECIFIC ORGANS FOLLOWING INTRAVENOUS AND INTRA ARTERIAL ADMINISTRATION IN BEAGLE DOGS
Kaske, M. et al.
J. Pharm. Sci., 69, 1980, p. 755

937
SUPERIOR SWELLING PROPERTIES OF RESINS OF POLY-N-ACRYLYL DIALKYL
AMINES OVER POLYSTYRENE IN SOLVENTS FOR PEPTIDE SYNTHESIS
Stahl, G.L.
Int. J. Pept. Protein Res., 15, 1980, p. 331

Styrene Sulphonate Polymer

938
INTERACTION OF NUCLEATED CELLS WITH ANIONIC POLYELECTROLYTES AND
INHIBITION OF THE INTERACTION BY PANCREATIC DEOXYRIBONUCLEASE
Tunis, M. & Regelson, W.
Exp. Cell. Res., 40, 1965, p. 383

939
SIMPLE METHOD FOR THE RAPID AND ECONOMICAL IMMOBILISATION OF
GLUCOSE ISOMERASE
Bhatt, R.R. et al.
Enzyme, Microb. Technol., 1, 1979, p. 113

Styrene-Acrylic Copolymer

940
MOLECULAR SCALE DRUG ENTRAPMENT AS A PRECISE METHOD OF CON-
TROLLED DRUG RELEASE. IV. ENTRAPMENT OF ANIONIC DRUGS BY POLYMERIC
GELATION
Boylan, J.C. & Banker, G.S.
J. Pharm. Sci., 62, 1973, p. 1177
PHENOBARBITAL

941
ANTIMICROBIAL POLYMERS
Ackart, W.B. et al.
J. Biomed. Mater. Res., 9, 1975, p. 55

Styrene-Glycidyl Methacrylate Copolymer

942
DRUG ENTRAPMENT FOR CONTROLLED RELEASE IN RADIATION-POLYMERISED
BEADS
Yoshida, M. et al.
J. Pharm. Sci., 68, 1979, p. 628
POTASSIUM CHLORIDE

Styrene-Isoprene Copolymer

943
SYNTHETIC POLYMERS WITH ANTICOAGULANT ACTIVITY
Van der Does, L. et al.
J. Polym. Sci. Polym. Symp., No. 66, 1979, p. 337

Styrene-Maleic Acid Copolymer

944
ENTERIC COATINGS. IV. IN VIVO TESTING OF GRANULES AND TABLETS
COATED WITH STYRENE-MALEIC ACID COPOLYMER
Wagner, J.G. et al.
J. Am. Pharm. Assn., 49, 1960, p. 128

945
SYNTHETIC ANTIGENS. IV. ANTIGEN-BINDING CELLS TO INTERPOLYMER OF
STYRENE AND MALEIC ACID IN RATS IMMUNISED WITH SHEEP RED BLOOD
CELLS
Skibinski, G. et al.
Arch. Immunol. Ther. Exp., 27, 1979, p. 615

Styrene-Maleic Anhydride Copolymer

946
INTERFERON-STIMULATING AND IN VIVO ANTIVIRAL EFFECTS OF VARIOUS
SYNTHETIC ANIONIC POLYMERS
Merigan, T.C. & Finkelstein, M.S.
Virology, 35, 1968, p. 363

Styrene-Vinyl Ether Pyridine Copolymer

947
DISINTEGRATION OF PROTECTIVE-COATED TABLETS AS DETERMINED BY
URINARY EXCRETION IN HUMANS
Ida, T. et al.
J. Pharm. Sci., 52, 1963, p. 472
RIBOFLAVIN

SULPHONAMIDE POLYMERS AND COPOLYMERS

Sulphonamide Polymer

948
PHARMACOLOGICALLY ACTIVE POLYMERS. XI. POLYMERIC SULPHONAMIDES AS
POTENTIAL ANTIBACTERIALS AND CARRIERS FOR ANTITUMOUR AGENTS
Hofmann, V. et al.
Makromol. Chem., 177, 1976, p. 1791

Sulphonamide-Dimethylolurea Copolymer

949
SYNTHETIC BIOLOGICALLY ACTIVE POLYMERS. VI. SULPHONAMIDE- AND
SULPHONE-DIMETHYLOLUREA COPOLYMERS
Dombroski, J.R. & Donaruma, L.G.
J. Appl. Polym. Sci., 15, 1971, p. 1219

950
SYNTHETIC BIOLOGICALLY ACTIVE POLYMERS. 8. ANTIBACTERIAL ACTIVITY OF
SOME SULPHONAMIDE-DIMETHYLOLUREA COPOLYMERS
Dombroski, J.R. & Donaruma, L.G.
J. Med. Chem., 14, 1971, p. 460

THIOSEMICARBAZIDE POLYMER

951
ANTIMICROBIAL AND ANTIPARASITIC POLYMERS
Brierly, J.A. et al.
ACS Coatings & Plastics Preprints, 42, 1980, p. 432

UREA FORMALDEHYDE POLYMER

952
NEW, HIGHLY EFFECTIVE BUT NON-TOXIC ANTIBACTERIAL SUBSTANCE
Haler, D. & Aebi, A.
Nature, 190, May 20, 1961, p. 734

953
POLYMERIC ANTIMICROBIAL DRUGS
Donaruma, L.G. et al.
ACS Polymer Preprints, 20, No. 1, 1979, p. 346

URETHANE POLYMERS

Diphenylmethane Diisocyanate Polymer

954
HYDROGELS FOR THE CONTROLLED RELEASE OF PROSTAGLANDIN E2
Graham, N.B. et al.
ACS Polym. Preprints, 21, 1980, p. 104
PESSARY

Ether Urethane Polymer

955
CHARACTERISATION OF CONTROLLED RELEASE OF PROSTAGLANDINS FROM
POLYMER MATRICES FOR THROMBUS PREVENTION
McRea, J.C. & Kim, S.W.
Trans. Am. Soc. Artif. Int. Org., 24, 1978, p. 746

956
PROGESTIN PERMEATION THROUGH POLYMER MEMBRANES. I. DIFFUSION
STUDIES ON PLASMA-SOAKED MEMBRANES
Zentner, G.M. et al.
J. Pharm. Sci., 67, 1978, p. 1347
PROGESTERONE

957
PROSTAGLANDIN RELEASING POLYMERS FROM NONTHROMBOGENIC
SURFACES
McRea, J.C. & Kim, S.W.
Plastics in Medicine and Surgery, June 1979, p. 31.1
Plastics and Rubber Institute

958
COMPARISON OF SEGMENTED POLYETHER URETHANE WITH POLYETHYLENE
IUDs IN RABBITS
Stolzenberg, S.J. et al.
Contraception, 20, 1979, p. 91

959
BIOPHYSICAL EVALUATION OF THE TUMORIGENIC RESPONSE TO IMPLANTED
POLYMERS
Habal, M.B. & Powell, R.D.
J. Biomed. Mater. Res., 14, 1980, p. 447

960
IMMUNOLOGICAL EVALUATION OF THE TUMORIGENIC RESPONSE TO
IMPLANTED POLYMERS
Habal, M.B.
J. Biomed. Mater. Res., 14, 1980, p. 455

Urethane Polymer

961
CANCER INDUCTION BY POLYURETHANE AND POLYSILICONE PLASTICS
Hueper, W.C.
J. Nat. Cancer Inst., 33, 1964, p. 1005

962
STUDIES ON MICROCAPSULES. I. PREPARATION OF POLYURETHANE AND
POLYPHENOLESTER MICROCAPSULES
Suzuki, S. et al.
Chem. Pharm. Bull., 16, 1968, p. 1629
SODIUM HYDROXIDE

963
STUDIES ON MICROCAPSULES. III. PERMEABILITY OF POLYURETHANE
MICROCAPSULE MEMBRANES
Shigeri, Y. & Kondo, T.
Chem. Pharm. Bull., 17, 1969, p. 1073

964
BIOLOGICAL ACTIVITY OF POLYMERS
Conning, D.M.
Plastics in Medicine & Surgery Symposium, 15 & 16 Sept. 1971

965
USE OF THERMAL GRAVIMETRIC ANALYSIS IN SORPTION STUDIES. II.
EVALUATION OF DIFFUSIVITY AND SOLUBILITY OF A SERIES OF ALIPHATIC
ALCOHOLS IN POLYURETHANE
Hung, G.W.C. & Autian, J.
J. Pharm. Sci., 61, 1972, p. 1094

966
BIOLOGICAL ACTIVITY OF IONENE POLYMERS
Rembaum, A.
Appl. Polym. Symp., No. 22, 1973, p. 299

967
SURFACE BONDED HEPARIN
Falb, R.D.
Polym. Sci. Technol., 8, 1974, p. 77
Johnson and Johnson Symposium, N.J. July, 1974

968
ENZYMES IMMOBILISED BY MICROENCAPSULATION WITHIN SPHERICAL ULTRATHIN POLYMERIC MEMBRANES
Chang, T.M.S.
J. Macromol. Sci. Chem., A10, 1976, p. 254
SEBACOYL CHLORIDE

969
DEVELOPMENT OF AN ESTRIOL-RELEASING INTRAUTERINE DEVICE
Baker, R.W. et al.
ACS Coatings and Plastics Preprints, 37, 1977, p. 421

970
PROGESTIN PERMEATION THROUGH POLYMER MEMBRANES. I. DIFFUSION STUDIES ON PLASMA-SOAKED MEMBRANES
Zentner, G.M. et al.
J. Pharm. Sci., 67, 1978, p. 1347
PROGESTERONE

971
DEVELOPMENT OF AN ESTRIOL-RELEASING INTRAUTERINE DEVICE
Baker, R.W. et al.
J. Pharm. Sci., 68, 1979, p. 20

972
CONTROLLED RELEASE MICROPUMP FOR INSULIN ADMINISTRATION
Sefton, M.V. et al.
Ann. Biomed. Engng., 7, 1979, p. 329

973
CONTROLLED RELEASE MICROPUMP FOR INSULIN DELIVERY AT VARIABLE RATES
Sefton, M.V. & Burns, K.J.
ACS Polym. Preprints, 21, 1980, p. 105

VINYL ACETATE POLYMERS AND COPOLYMERS

Vinyl Acetate Polymer

974
EVALUATION OF POLYMERIC MATERIALS. I. SCREENING OF SELECTED POLYMERS AS FILM COATING AGENTS
Munden, B.J. et al.
J. Pharm. Sci., 53, 1964, p. 395

975
EVALUATION OF POLYMERIC MATERIALS. II. SCREENING OF SELECTED VINYLS AND ACRYLATES AS PROLONGED-ACTION COATINGS
Nessel, R.J. et al.
J. Pharm. Sci., 53, 1964, p. 790
AMPHETAMINE SULPHATE

976
KINETICS OF EFFECT LEVELS AFTER HYOSCYAMINE IN SUSTAINED RELEASE TABLETS
Brandstrom, A. & Sjogren, J.
Acta Pharm. Suec., 4, 1967, p. 157

977
LITHIUM ABSORPTION FROM SUSTAINED RELEASE TABLETS
Amdisen, A. & Sjogren, J.
Acta Pharm. Suec., 5, 1968, p. 465
LITHIUM SULPHATE

978
BEAD POLYMERISATION TECHNIQUE FOR SUSTAINED-RELEASE DOSAGE FORM
Khanna, S.C. et al.
J. Pharm. Sci., 59, 1970, p. 614
CHORAMPHENICOL; METHOXYPYRIDINYL SULPHANILAMIDE; CHLOROTHIAZIDE; PENTOBARBITAL; PAPERAVINE; HYDROCORTISONE; ETHYLPHENYLGLUTARIMIDE

979
POTASSIUM ABSORPTION FROM SUSTAINED RELEASE TABLETS
Graffner, C. & Sjogren, J.
Acta Pharm. Suec., 8, 1971, p. 13

980
SIDE EFFECTS OF POTASSIUM CHLORIDE IN PRODUCTS WITH DIFFERENT DISSOLUTION RATES
Graffner, C. & Sjogren, J.
Acta Pharm. Suec., 8, 1971, p. 19

981
ABSORPTION OF ALPRENOLOL IN MAN FROM TABLETS WITH DIFFERENT RATES OF RELEASE
Johansson R. et al.
Acta Pharm. Suec., 8, 1971, p. 59

982
EFFECTS OF POTASSIUM TABLETS OF DIFFERENT DISSOLUTION RATES ON MOTILITY PATTERN OF SMALL INTESTINE IN THE DOG
Sundell, G.
Acta Pharm. Suec., 8, 1971, p. 73
POTASSIUM CHLORIDE

983
POLYMERS CONTAINING PHENETHYLAMINES
Weiner, B.Z. et al.
J. Med. Chem., 15, 1972, p. 410

984
STRENGTH OF THE INSOLUBLE RESIDUES OF PLASTIC MATRIX SLOW RELEASE TABLETS IN VITRO AND IN VIVO
Dahlinder, L.K. et al.
Acta Pharm. Suec., 8, 1971, p. 323
MANNITOL

985
IN VITRO AND IN VIVO EVALUATION OF SUSTAINED RELEASE TABLETS CONTAINING NOREPHEDRINE CHLORIDE
Lindberg, N.O. & Persson, G.G.A.
Acta Pharm. Suec., 9, 1972, p. 237

986
STRENGTH OF THE INSOLUBLE RESIDUES OF PLASTIC MATRIX SLOW RELEASE TABLETS IN VITRO AND IN VIVO
Dahlinder, L.E. et al.
Acta Pharm. Suec., 10, 1973, p. 323
MANNITOL

987
DRUG RELEASE FROM HYDROXYPROPYL CELLULOSE-PVAC FILMS
Borodkin, S. & Tucker, F.E.
J. Pharm. Sci., 63, 1974, p. 1359
METHAPYRILENE; PENTOBARBITAL; SALICYLIC ACID

988
RELEASE STUDIES OF NITROFURANTOIN FROM POLYVINYLACETATE-CROTONIC ACID COPOLYMERS
El-Khouly, A.E. et al.
Canad. J. Pharm. Sci., 9, 1974, p. 54

989
COMPARATIVE ABSORPTION AND PHARMOKINETIC STUDIES ON PROXYPHYLLINE IN ORDINARY AND SLOW RELEASE TABLETS
Graffner, C. et al.
Acta Pharm. Suec., 11, 1974, p. 125

990
UNIFIED MATHEMATICAL MODEL FOR DIFFUSION FROM DRUG-POLYMER COMPOSITE TABLETS
Fu, J.C. et al.
J. Biomed. Mater. Res., 10, 1976, p. 743
HYDROCORTISONE

991
MATHEMATICAL MODELS FOR THE RELEASE OF DRUGS FROM MATRIX TABLETS
Cobby, J.
J. Biomed. Mater. Res., 12, 1978, p. 627
HYDROCORTISONE

992
POLYMERIC AFFINITY DRUGS FOR CARDIOVASCULAR, CANCER AND UROLITHIASIS THERAPY
Goldberg, E.P.
Polymer Drugs, 1978, p. 239, Academic Press
Donaruma, L.G. & Vogl, O. eds.

993
STABILISATION OF SULPHISOMIDINE TABLETS BY USE OF FILM COATING CONTAINING UV ABSORBER: PROTECTION OF COLOURATION AND PHOTOLYTIC DEGRADATION FROM EXAGGERATED LIGHT
Matsuda, Y. et al.
J. Pharm. Sci., 67, 1978, p. 196

994
RECENT PROBLEMS CONCERNING FUNCTIONAL MONOMERS AND POLYMERS
CONTAINING NUCLEIC ACID BASES
Takemoto, K.
Polymeric Drugs, 1978, p. 103, Academic Press
Donaruma, L.G. & Vogl, O. eds.

995
CONTROLLED SLOW RELEASE OF CHEMOTHERAPEUTIC DRUGS FOR CANCER
FROM MATRICES PREPARED BY RADIATION POLYMERISATION AT LOW
TEMPERATURES
Kaetsu, I.
J. Biomed. Mat. Res., 14, 1980, p. 185
MITOMYCIN; 5-FLUOROURACIL; BLEOMYCIN

996
SOME NEW REACTIVE POLYMERS FOR THE IMMOBILISATION OF ENZYMES
Manecke, G. & Vogt, H.G.
Biochimie, 62, 1980, p. 603
PAPAIN; TRYPSIN; CHYMOTRYPSIN; UREASE; GLUCOSE OXDASE; CATALASE;
GLUCOSE-6-PHOSPHATE DEHYDROGENASE; TRANSFERASE HEXOKINASE

Vinyl Acetate-Maleic Anhydride Copolymer

997
INTERFERON-STIMULATING AND IN VIVO ANTIVIRAL EFFECTS OF VARIOUS
SYNTHETIC ANIONIC POLYMERS
Merigan, T.C. & Finkelstein, M.S.
Virology, 35, 1968, p. 363

998
CONTROLLED DRUG RELEASE BY POLYMER DISSOLUTION. I. PARTIAL ESTERS OF
MALEIC ANHYDRIDE COPOLYMERS — PROPERTIES AND THEORY
Heller, J. et al.
J. Appl. Polym. Sci., 22, 1978, p. 1991
HYDROCORTISONE

Vinyl Acetate-Vinyl Chloride Copolymer

999
EFFECT OF SINTERING ON THE PORE STRUCTURE AND STRENGTH OF PLASTIC
MATRIX TABLET
Rowe, R.C. et al.
J. Pharm. Pharmacol., 25, 1973, p. 12P
POTASSIUM CHLORIDE

VINYL ALCOHOL POLYMER

1000
EFFECTS OF INTRAVENOUS AND INTRAPERITONEAL INTRODUCTION OF POLYVINYL ALCOHOL SOLUTION UPON THE BLOOD
Hueper, W.C. et al.
J. Pharm. Exp. Ther., 70, 1940, p. 201
OSMOTIC BALANCE

1001
EFFECT OF METHYL CELLULOSE ON RESPONSES TO SOLUTIONS OF PILOCAR-PINE
Haas, J.S. & Merrill, D.L.
Amer. J. Ophthalmol., 54, 1962, p. 21

1002
INTERACTION BETWEEN POLYVINYL ALCOHOL AND PROCAINE HYDROCHLORIDE
Botre, C. & Riccieri, F.M.
J. Pharm. Sci., 52, 1963, p. 1011

1003
POLYVINYL ALCOHOL AS AN OPHTHALMIC VEHICLE. EFFECT ON REGENERATION OF CORNEAL EPITHELIUM
Krishna, N. & Brow, F.
Amer. J. Ophthalmol., 57, 1964, p. 99

1004
STUDY OF THE POLYVINYL ALCOHOL-BORATE-IODINE COMPLEX. III. DETECTION OF BORATES IN URINE
Monte-Bovi, A.J. et al.
J. Pharm. Sci., 53, 1964, p. 1278

1005
EVALUATION OF POLYMERIC MATERIALS. IV. GRANULATING AGENTS FOR COMPRESSED TABLETS
Willis, C.R. et al.
J. Pharm. Sci., 54, 1965, p. 366

1006
HYPERTENSION FOLLOWING SUBCUTANEOUS AND INTRAPERITONEAL INJECTIONS OF POLYLVINYL ALCOHOL AND THE EFFECT OF ALDOSTERONE
Hall, C.E. & Hall, O.
Can. J. Physiol. Pharmacol., 45, 1967, p. 161

1007
IMMUNOLOGICAL PROPERTIES OF AMYLOSE, DEXTRAN AND POLYVINLALCOHOL CONJUGATED WITH POLYTYROSYL-PEPTIDES
Sorg, C. et al.
Eur. J. Biochem., 17, 1970, p. 85

1008
POLYVINYL ALCOHOL-HEPARIN HYDROGEL
Merrill, E.W. et al.
J. Appl. Physiol., 29, 1970, p. 723

1009
INTERACTION OF DRUGS WITH POLYMERS. III. PHASE SEPARATION OF POLYACIDS BY O-BENZOYLTHIAMINE DISULPHIDE HYDROCHLORIDE AND GASTROINTESTINAL ABSORPTION OF O-BENZOYLTHIAMINE DISULPHIDE-POLYACID COMPLEXES
Tanaka, N. et al.
Chem. Pharm. Bull., 18, 1970, p. 1083

1010
OCULAR EVALUATION OF POLYVINYL ALCOHOL VEHICLE IN RABBITS
Patton, T.F. & Robinson, J.R.
J. Pharm. Sci., 64, 1975, p. 1312
PILOCARPINE

1011
PHARMACOLOGICAL STUDIES WITH SLOW-RELEASE FORMULATIONS OF OXPRENOLOL IN MAN
Davidson, C. et al.
Eur. J. Clin. Pharmacol., 10, 1976, p. 189

1012
SUSTAINED RELEASE OF MACROMOLECULES FROM POLYMERS
Langer, R. & Folkman, J.
Midland Macromol. Inst., Polymeric Delivery Systems, 5th Int. Symp., Aug. 1976, p. 175
SOYBEAN TRYPSIN INHIBITOR; LYSOZYME; ALKALINE PHOSPHATASE; CATALASE; INSULIN; HEPARIN

1013
POLYMERS FOR THE SUSTAINED RELEASE OF PROTEINS AND OTHER
MACROMOLECULES
Langer, R. & Folkman, J.
Nature, 263, Oct. 28, 1976, p. 797

1014
USE OF POLYMERS FOR THE SUSTAINED RELEASE OF PROTEINS AND OTHER
LARGE MOLECULES
Langer, R. & Folkman, J.
ACS Polymer Preprints, 18, No. 2, 1977, p. 379
LYSOZYME; ALKALINE PHOSPHATASE

1015
DEVELOPMENT OF SEMICRYSTALLINE PVAL HYDROGELS FOR BIOMEDICAL
APPLICATIONS
Peppas, N.A. & Merrill, E.W.
J. Biomed. Mater. Res., 11, 1977, p. 423
HEPARIN

1016
MICROENCAPSULATION AND FORMULATION OF SUSTAINED RELEASE
CAPSULES OF DIAZEPAM
Shukla, A.K. & Sharma, S.N.
Ind. J. Pharm., 39, 1977, p. 100

1017
MICROENCAPSULATION OF PHENOBARBITAL BY SPRAY POLYCONDENSATION
Voellmy, C. et al.
J. Pharm. Sci., 66, 1977, p. 631

1018
POLYMERIC AFFINITY DRUGS FOR CARDIOVASCULAR CANCER AND
UROLITHIASIS THERAPY
Goldberg, E.P.
Polymeric Drugs, 1978, p. 239, Academic Press
Donaruma, L.G. & Vogl, O. eds.

1019
IMMOBILISED ENZMES
Manecke, G. & Schlunsen, J.
Polymeric Drugs, 1978, p. 39, Academic Press
Donaruma, L.G. & Vogl, O. eds.

1020
SYNTHESIS AND CHARACTERISATION OF POLYMERIC DERIVATIVES OF THE ANTITUMOUR AGENT METHOTREXATE
Przybylski, M. et al.
Makromol. Chem., 179, 1978, p. 1719

1021
POLYMERIC DRUGS IN THE CHEMOTHERAPY OF MICROBIAL INFECTIONS
Samour, C.M.
Polymeric Drugs, 1978, p. 161, Academic Press
Donaruma, L.G. & Vogl, O. eds.

1022
CONTROLLED CHEMOTHERAPY THROUGH MACROMOLECULES
Zaffaroni, A. & Bonson, P.
Polymeric Drugs, 1978, p. 1, Academic Press
Donaruma, L.G. & Vogl, O. eds.

1023
POLYVINYL ALCOHOL ENCAPSULATED ACTIVATED CHARCOAL AS A TOXIN ADSORBENT
Gopinathan, C. et al.
Symposium on Industrial Polymers & Radiation
Gujarat, Feb. 1979

1024
POLYVINYL ALCOHOL HYDROGELS AS A DRUG CARRIER
Yamauchi, A. et al.
ACS Polymer Preprints, 20, No. 1, 1979, p. 575
PILOCARPINE

1025
ANTIMICROBIAL AND ANTIPARASITIC POLYMERS
Brierley, J.A. et al.
ACS Coatings & Plastics Preprints, 42, 1980, p. 432

1026
SOME NEW REACTIVE POLYMERS FOR THE IMMOBILISATION OF ENZYMES
Manecke, G. & Vogt, H.G.
Biochimie, 62, 1980, p. 603
PAPAIN; TRYPSIN; CHYMOTRYPSIN; UREASE; GLUCOSE OXIDASE; CATALASE;
GLUCOSE-6-PHOSPHATE DEHYDROGENASE; TRANSFERASE; HEXOKINASE

VINYL CHLORIDE POLYMERS AND COPOLYMERS

Vinyl Chloride Polymer

1027
FURTHER STUDIES OF POLYMERS AS CARCINOGENIC AGENTS IN ANIMALS
Oppenheimer, B.S. et al.
Cancer Res., 15, 1955, p. 333

1028
EVALUATION OF POLYMERIC MATERIALS. I. SCREENING OF SELECTED POLYMERS AS FILM COATING AGENTS
Munden, B.J. et al.
J. Pharm. Sci., 53, 1964, p. 395

1029
EVALUATION OF POLYMERIC MATERIALS. II. SCREENING OF SELECTED VINYLS AND ACRYLATES AS PROLONGED-ACTION COATINGS
Nessel, R.J. et al.
J. Pharm. Sci., 53, 1964, p. 790
AMPHETAMINE SULPHATE

1030
SUSTAINED-RELEASE ASPIRIN TABLET USING AN INSOLUBLE MATRIX
Vora, M.S. et al.
J. Pharm. Sci., 53, 1964, p. 487

1031
INVESTIGATION OF FACTORS INFLUENCING RELEASE OF SOLID DRUG DISPERSED IN INERT MATRICES. IV
Desai, S.J. et al.
J. Pharm. Sci., 55, 1966, p. 1235

1032
RELEASE RATES OF SOLID DRUG MIXTURES DISPERSED IN INERT MATRICES. I. NONINTERACTING DRUG MIXTURES
Singh, P. et al.
J. Pharm. Sci., 56, 1967, p. 1542
SALICYLIC ACID; BENZOIC ACID

1033
RELEASE RATES OF SOLID DRUG MIXTURES DISPERSED IN INERT MATRICES. II.
MUTUALLY INTERACTING DRUG MIXTURES
Singh, P. et al.
J. Pharm. Sci., 56, 1967, p. 1548
BENZOCAINE; CAFFEINE

1034
TOXICITY PROFILES OF VINYL AND POLYOLEFINIC PLASTICS AND THEIR
ADDITIVES
Guess, W.L. & Huberman, S.
Biomed. Mat. Res., 2, 1968, p. 313

1035
EXPERIMENTAL ERRORS RESULTING FROM UPTAKE OF LIPOPHILIC DRUGS BY
SOFT PLASTIC MATERIALS
Minder, R. et al.
Biochem. Pharmacol., 19, 1970, p. 2179
IMIPRAMINE

1036
BIOLOGICAL ACTIVITY OF POLYMERS
Conning, D.M.
Plastics in Medicine & Surgery Symposium, 15 & 16, Sept. 1971

1037
SUSTAINED RELEASE FORMULATION OF PREDNISOLONE ADMINISTERED
ORALLY TO MAN
D'Arcy, P.F. et al.
J. Pharm. Sci., 60, 1971, p. 1028

1038
STRENGTH OF THE INSOLUBLE RESIDUES OF PLASTIC MATRIX SLOW RELEASE
TABLETS IN VITRO AND IN VIVO
Dahlinder, L.K. et al.
Acta Pharm. Suec., 8, 1971, p. 323
MANNITOL

1039
HOT EXTRUDED DOSAGE FORMS. I. TECHNOLOGY AND DISSOLUTION KINETICS
OF POLYMERIC MATRICES
El-Egakey, M.A. et al.
Pharm. Acta Helv., 46, 1971, p. 31

1040
STUDIES ON SUSTAINED RELEASE FORMULATIONS OF TOLBUTAMIDE
Bhoyar, G.S. et al.
Ind. J. Pharmacy, 34, 1972, p. 151

1041
STRENGTH OF THE INSOLUBLE RESIDUES OF PLASTIC MATRIX SLOW RELEASE
TABLETS IN VITRO AND IN VIVO
Dahlinder, L.K. et al.
Acta Pharm. Suec., 10, 1973, p. 323
MANNITOL

1042
SAFETY TESTING OF MEDICAL PLASTICS. I. ASSESSMENT OF METHODS
Pelling, D. et al.
Food Cosmet. Toxicol., 11, 1973, p. 69

1043
SURFACE BONDED HEPARIN
Falb, R.D.
Polym. Sci. Technol., 8, 1974, p. 77
Johnson and Johnson Symposium, N.J., July 1974

1044
SUSTAINED RELEASE CHLORPHENIRAMINE MALEATE TABLETS
Kassem, A.A. et al.
Bull. Fac. Pharmacy, 13, 1974, p. 37

1045
ANTIMICROBIAL POLYMERS
Ackart, W.B. et al.
J. Biomed. Mater. Res., 9, 1975, p. 55
BENZALKONIUM SALTS; 8-HYDROXYQUINOLINIUM

1046
CHARACTERISATION OF CONTROLLED RELEASE OF PROSTAGLANDINS FROM
POLYMER MATRICES FOR THROMBUS PREVENTION
McRea, J.C. & Kim, S.W.
Trans. Am. Soc. Artif. Int. Org., 24, 1978, p. 746

1047
SYNTHESIS AND SOME PROPERTIES OF ANTITHROMBOGENIC POLYMERS
Plate, N.A.
Polymeric Drugs, 1978, p. 63, Academic Press
Donaruma, L.G. & Vogl, O. eds.

1048
PROSTAGLANDIN RELEASING POLYMERS FROM NONTHROMBOGENIC SURFACES
McRea, J.C. & Kim, S.W.
Plastics in Medicine and Surgery, June 1979, p. 31.1
Plastics and Rubber Institute

1049
ANTIMICROBIAL AND ANTIPARASITIC POLYMERS
Brierley, J.A. et al.
ACS Coatings & Plastics Preprints, 42, 1980, p. 432

1050
IMMOBILISATION OF ENZYMES FOR MEDICAL USES ON PLASTIC SURFACES BY RADIATION-INDUCED POLYMERISATION AT LOW TEMPERATURES
Kaetsu, I.
J. Biomed. Mat. Res., 14, 1980, p. 199
GLUCOSE OXIDASE

Vinyl Chloride-Vinyl Acetate Copolymer

1051
SOLID DOSAGE FORM WITH CONSTANT RELEASE
Lerk, C.F. et al.
J. Pharm. Pharmacol., 25, 1973, (Suppl.), p. 151P
POTASSIUM CHLORIDE; ORPHENADRINE

VINYL ETHER POLYMERS AND COPOLYMERS

Chloroethyl Vinyl Ether Polymer

1052
IMMOBILISATION OF ENZYMES WITH USE OF PHOTOSENSITIVE POLYMERS
HAVING THE STILBAZOLIUM GROUP
Ichimura, K. & Watanabe, S.
J. Polym. Sci. Polym. Chem., 18, 1980, p. 891
INVERTASE; GLUCOAMYLASE; CATALASE

Divinyl Ether-Maleic Anhydride Copolymer

1053
ETHYLENE MALEIC ANHYDRIDE COPOLYMERS AS VIRAL INHIBITORS
Felz, E.T. & Regelson, W.
Nature, 196, 17 Nov. 1962, p. 642

1054
INDUCTION OF CIRCULATING INTERFERON BY SYNTHETIC ANIONIC POLYMERS
OF KNOWN COMPOSITION
Merigan, T.C.
Nature, 214, 22 Apr. 1967, p. 416

1055
INTERFERON INDUCTION IN MAN BY A SYNTHETIC POLYANION OF DEFINED
COMPOSITION
Merigan, T.C. & Regelson, W.
N. Engl. J. Med., 277, 1967, p. 1283

1056
PREVENTION AND TREATMENT OF FRIEND LEUKEMIA VIRUS INFECTION BY
INTERFERON-INDUCING SYNTHETIC POLYANIONS
Regelson, W.
Adv. Expl. Med. Biol., 1, 1967, p. 315

1057
INTERFERON-STIMULATING AND IN VIVO ANTIVIRAL EFFECTS OF VARIOUS
SYNTHETIC ANIONIC POLYMERS
Merigan, T.C. & Finkelstein, M.S.
Virology, 35, 1968, p. 363

1058
LOCAL AND SYSTEMIC PROTECTION BY SYNTHETIC POLYANIONIC INTERFERON
INDUCERS IN MICE AGAINST INTRANASAL VESICULAR STOMATITIS VIRUS
De Clercq, E. & Merigan, T.C.
J. Gen. Virol., 5, 1969, p. 359

1059
PROLONGED ANTIVIRAL PROTECTION BY INTERFERON INDUCERS
De Clercq. E. & De Somer, P.
Proc. Soc. Exp. Biol. & Med., 132, 1969, p. 699

1060
INHIBITION OF ADJUVANT DISEASE IN RATS BY THE INTERFERON-INDUCING
AGENT PYRAN COPOLYMER
Kapusta, M.A. & Mendelson, J.
Arthritis. Rheum., 12, 1969, p. 463

1061
COAGULATION STUDIES WITH LINEAR COPOLYMERS OF ALIPHATIC
HYDROCARBONS AND MALEIC ACID: A NEW CLASS OF ANTICOAGULANTS
Shamash, Y. & Alexander, B.
Biochim. Biophys. Acta., 194, 1969, p. 449

1062
STIMULATION OF ANTIBODY FORMATION BY PYRAN COPOLYMER
Braun, W. et al.
Proc. Soc. Exp. Biol. & Med., 133, 1970, p. 171

1063
IMMUNOSUPPRESSION, INTERFERON INDUCERS AND LEUKEMIA IN MICE
Hirsch, M.S. et al.
Proc. Soc. Exp. Biol. & Med., 134, 1970, p. 309

1064
CLINICAL STUDIES EMPLOYING INTERFERON INDUCERS IN MAN AND ANIMALS
Merigan, T.C. et al.
Ann. N.Y. Acad. Sci., 173, 1970, p. 746

1065
BIPHASIC RESPONSE OF THE RETICULOENDOTHELIAL SYSTEM INDUCED BY
PYRAN COPOLYMER
Munson, A.E. et al.
RES J. Reticuloendothielial Soc., 7, 1970, p. 375

1066
PROTECTION OF MICE AGAINST BACTERIAL INFECTION BY INTERFERON INDUCERS
Pindak, F.F.
Infect. Immun., 1, 1970, p. 271

1067
RETICULOENDOTHELIAL EFFECTS OF INTERFERON INDUCERS: POLYANIONIC AND NON-POLYANIONIC PHYLAXIS AGAINST MICROORGANISMS
Regelson, W. & Munson, A.E.
Ann. N.Y. Acad. Sci., 173, 1970, p. 831

1068
SYNTHETIC POLYANIONS PROTECT MICE AGAINST INTRACELLULAR BACTERIAL INFECTION
Remington, J.S. & Merigan, T.C.
Nature, 226, 25 Apr. 1970, p. 361

1069
EFFECT OF PYRAN COPOLYMER ON PHAGOCYTOSIS AND TUMOUR GROWTH
Kapila, K. et al.
RES. J. Reticuloendothelial Soc., 9, 1971, p. 447

1070
SENSITISATION TO ENDOTOXIN BY PYRAN COPOLYMER
Munson, A.E. & Regelson, W.
Proc. Soc. Exp. Biol. & Med., 137, 1971, p. 553

1071
DISTRIBUTION AND EXCRETION OF C-14 TAGGED PYRAN COPOLYMER
Regelson, W. et al.
RES. J. Reticuloendothelial Soc., 6, 1971, p. 313

1072
INTERFERON INDUCERS: ENHANCEMENT OF VIRAL ONCOGENESIS IN MICE AND RATS
Gazdar, A.F.
Proc. Soc. Exp. Biol. Med., 139, 1972, p. 1132

1073
EFFECT OF INTERFERON INDUCERS AND INTERFERON ON BACTERIAL INFECTIONS
Giron, D.J. et al.
Antimicrob. Agents Chemother., 1, 1972, p. 80

1074
EFFECTS OF PYRAN COPOLYMER ON ONCOGENIC VIRUS INFECTIONS IN IMMUNOSUPPRESSED HOSTS
Hirsch, M.S. et al.
J. Immunol., 108, 1972, p. 1312

1075
PYRAN AND POLYRIBONUCLEOTIDES: DIFFERENCES IN BIOLOGICAL ACTIVITIES
Morahan, P.S. et al.
Antimicrob. Agents Chemother., 2, 1972, p. 16

1076
DIVINYL ETHER-MALEIC ANHYDRIDE (PYRAN) COPOLYMER: THE EFFECT OF MOLECULAR WEIGHT ON BIOLOGICAL ACTIVITY
Breslow, D.S. et al.
Nature, 246, 1973, p. 160

1077
EFFECTS OF PYRAN COPOLYMER ON LEUKEMOGENESIS IN IMMUNOSUPPRESSED AKR MICE
Hirsch, M.S. et al.
J. Immunol., 111, 1973, p. 91

1078
DETERMINATION OF FIBRINOGEN IN NORMAL HUMAN PLASMA IN THE PRESENCE AND ABSENCE OF ANTICOAGULANTS
Roberts, P.S. et al.
J. Lab. Clin. Med., 82, 1973, p. 822

1079
INDUCTION OF MACROPHAGE-MEDIATED TUMOUR CELL CYTOTOXICITY BY PYRAN COPOLYMER
Kaplan, A.M. et al.
J. Natl. Cancer Inst., 52, 1974, p. 1919

1080
ACCELERATED DEVELOPMENT OF BENZOPYRENE-INDUCED SKIN TUMOURS IN MICE TREATED WITH PYRAN COPOLYMER
Kripke, M.L. & Borsos, T.
J. Nat. Cancer Inst., 53, 1974, p. 1409

1081
ANTITUMOUR ACTION OF PYRAN COPOLYMER AND TILORONE AGAINST LEWIS LUNG CARCINOMA AND B-16 MELANOMA
Morahan, P.S. et al.
Cancer Res., 34, 1974, p. 506

1082
INHIBITION OF RNA DEPENDENT DNA POLYMERASE OF AVIAN
MYELOBLASTOSIS VIRUS BY PYRAN COPOLYMER
Papas, T.S. et al.
Proc. Natl. Acad..Sci., USA, 71, 1974, p. 367

1083
COMBINED DRUG AND IMMUNOSTIMULATION THERAPY AGAINST A
SYNGENEIC MURINE LEUKEMIA
Pearson, J.W. et al.
J. Nat. Cancer Inst., 52, 1974, p. 463

1084
IMMUNOADJUVANT ACTIVITY OF PYRAN COPOLYMER
Baird, L.G. & Kaplan, A.M.
Cell. Immunol., 20, 1975, p. 167

1085
TUMOUR-CELL CYTOTOXICITY VERSUS CYTOSTASIS OF PYRAN ACTIVATED
MACROPHAGES
Kaplan, A.M. et al.
Fogarty International Center Proceedings, 28, 1975

1086
PYRAN COPOLYMER AS AN EFFECTIVE ADJUVANT TO CHEMOTHERAPY
AGAINST A MURINE LEUKEMIA AND SOLID TUMOUR
Mohr, S.J. et al.
Cancer Res., 35, 1975, p. 3750

1087
INHIBITION AND ENHANCEMENT OF FRIEND LEUKEMIA VIRUS BY PYRAN
COPOLYMER
Schuller, G.B. et al.
Cancer Res., 35, 1975, p. 1915

1088
HISTOPATHOLOGY OF HOST RESPONSE TO LEWIS LUNG CARCINOMA:
MODULATION BY PYRAN
Snodgrass, M.J. et al.
J. Nat. Cancer Inst., 55, 1975, p. 455

1089
BIOLOGICALLY ACTIVE SYNTHETIC POLYMERS
Breslow, D.S.
Pure & Appl. Chem., 46, 1976, p. 103

1090
MACROPHAGE MEDIATED TUMOUR CELL CYTOTOXICITY
Kaplan, A.M. & Morahan, P.S.
Ann. N.Y. Acad. Sci., 276, 1976, p. 134

1091
PARADOXICAL EFFECTS OF IMMUNOPOTENTIATORS ON TUMOURS AND
TUMOUR VIRUSES
Morahan, P.S. et al.
J. Infect. Diseases, 133, Suppl., 1976, p. A249

1092
MACROPHAGE ACTIVATION AND ANTI-TUMOUR ACTIVITY OF BIOLOGIC AND
SYNTHETIC AGENTS
Morahan, P.S. & Kaplan, A.M.
Int. J. Cancer, 17, 1976, p. 82

1093
BIOLOGICAL ACTIVITY OF ANIONIC POLYMERS
Ottenbrite, R.M. et al.
ACS Polymer Preprints, 18, No. 1, 1977, p. 581

1094
COMPARATIVE STUDY OF ANTITUMOUR AND TOXICOLOGIC PROPERTIES OF
RELATED POLYANIONS
Ottenbrite, R. et al.
Polymer, 18, 1977, p. 461

1095
INTERFERON: AN INDUCER OF MACROPHAGE ACTIVATION BY POLYANION
Schultz, R.M. et al.
Science, 197, 1977, p. 674

1096
PHYSICO CHEMICAL CHARACTERISATION OF PYRAN COPOLYMER – AN
IMMUNOGENIC DRUG
Levine, H.I. et al.
ACS Polymer Preprints, 19, No. 2, 1978, p. 570

1097
POLYMERS AS INTERFERON INDUCERS
Levy, H.B.
Polymeric Drugs, 1978, p. 305, Academic Press
Donaruma, L.G. & Vogl, O. eds.

1098
BIOLOGICAL ACTIVITY OF POLYCARBOXYLIC ACID POLYMERS
Ottenbrite, R.M. et al.
Polymeric Drugs, 1978, p. 263, Academic Press
Donaruma, L.G. & Vogl, O. eds.

1099
SYNTHESIS AND CHARACTERISATION OF POLYMERIC DERVIATIVES OF THE
ANTITUMOUR AGENT METHOTREXATE
Przybylski, M. et al.
Makromol. Chem., 179, 1978, p. 1719

1100
POLYMERIC DERIVATIVES OF ACTIVATED CYCLOPHOSPHAMIDE AS DRUG
DELIVERY SYSTEMS IN ANTITUMOUR CHEMOTHERAPY
Hirano, T. et al.
Makromol. Chem., 180, 1979, p. 1125

1101
ACTIVATION OF MOUSE MACROPHAGES BY PYRAN COPOLYMER AND ROLE IN
AUGMENTATION OF NATURAL KILLER ACTIVITY
Puccetti, P. et al.
Int. J. Cancer, 24, 1979, p. 819

1102
BIOLOGICAL ACTIVITY OF POLYANIONS: PAST HISTORY AND NEW
PROSPECTIVES
Regelson, W.
J. Polym. Sci. Polym. Symp., No. 66, 1979, p. 483

1103
SYNTHESIS AND PROPERTIES OF ALTERNATING COPOLYMERS OF POTENTIAL
ANTITUMOUR ACTIVITY CONTAINING 5-FLUOROURACIL
Umrigar, P.P. et al.
J. Polym. Sci. Polym. Chem., 17, 1979, p. 351

1104
EFFECTS OF POLYANION IMMUNOMODULATORS ON THE IMMUNE SYSTEM
Baird, L.G. & Kaplan, A.M.
Anionic Polymeric Drugs, 1980, p. 185
Donaruma, L.G. et al., eds., Wiley-Interscience

1105
EFFECTS OF ANIONIC POLYMERIC DRUGS AND OTHER IMMUNOACTIVE
AGENTS ON HEPATIC MICROSOMAL MIXED-FUNCTION OXIDASES
Barnes, D.W.
Anionic Polymeric Drugs, 1980, p. 255
Donaruma, L.G. et al., eds., Wiley-Interscience

1106
ENHANCED MACROPHAGE AND NATURAL KILLER CELL ANTITUMOUR ACTIVITY
BY VARIOUS MOLECULAR WEIGHT MALEIC ANHYDRIDE DIVINYL ETHERS
Bartocci, A. et al.
J. Immunopharmacol., 2, 1980, p. 149

1107
ANTIVIRAL ACTIVITY OF SYNTHETIC POLYANIONS
Breinig, M.C. et al.
Anionic Polymeric Drugs, 1980, p. 211
Donaruma, L.G. et al., eds., Wiley-Interscience

1108
DIVINYL ETHER-MALEIC ANHYDRIDE COPOLYMER: ITS STRUCTURE AND
BIOLOGICAL PROPERTIES
Breslow, D.S.
ACS Polym. Preprints, 21, 1980, p. 1

1109
SYNTHESIS, CHARACTERISATION, AND BIOLOGICAL ACTIVITY OF PYRAN
COPOLYMERS
Butler, G.B.
Anionic Polymeric Drugs, 1980, p. 49
Donaruma, L.G. et al., eds., Wiley-Interscience

1110
PHYSICOCHEMICAL CHARACTERISTICS AND MOLECULAR PHARMACOLOGY OF
PYRAN COPOLYMER
Fiel, R.J. et al.
Anionic Polymeric Drugs, 1980, p. 143
Donaruma, L.G. et al., eds., Wiley-Interscience

1111
ANTITUMOUR ACTIVITY OF SYNTHETIC POLYANIONS
Kaplan, A.M.
Anionic Polymeric Drugs, 1980, p. 227
Donaruma, L.G. et al., eds., Wiley-Interscience

1112
CHANGES IN MACROPHAGE ECTOENZYMES ASSOCIATED WITH ANTI-TUMOUR ACTIVITY
Morahan, P.S. et al.
J. Immunol., 125, 1980, p. 1312

1113
STRUCTURE AND BIOLOGICAL ACTIVITIES OF SOME POLYANIONIC POLYMERS
Ottenbrite, R.M.
Anionic Polymeric Drugs, 1980, p. 21
Donaruma, L.G. et al., eds., Wiley-Interscience

1114
MACROPHAGE ACTIVATION AND MOBILISATION IN NUDE MICE BY CORYNEBACTERIUM PARVUM AND PYRAN: A FUNCTIONAL AND HISTOLOGIC STUDY
Stinnett, S.D. & Majeski, J.A.
J. Surg. Oncol., 14, 1980, p. 327

1115
EFFECT OF PYRAN COPOLMER ON MURINE HEMOPOIESIS
Zander, A.R. et al.
Exp. Hematol., 8, 1980, p. 521

1116
PYRAN COPOLYMER: EFFECT OF MOLECULAR WEIGHT ON STEM CELL MOBILISATION IN MICE
Zander, A.R. et al.
Biomedicine, 33, 1980, p. 69

Isobutyl Vinyl Ether-Maleic Anhydride Copolymer

1117
DISSOLUTION OF ALKYL VINYL ETHER-MALEIC ANHYDRIDE COPOLYMERS AND ESTER DERIVATIVES
Woodruff, C.W. et al.
J. Pharm. Sci., 61, 1972, p. 1916

Methyl Vinyl Ether-Maleic Acid Copolymer

1118
SYNTHETIC POLYMERS AS POTENTIAL SUSTAINED-RELEASE COATINGS
Kleber, J.W. et al.
J. Pharm. Sci., 53, 1964, p. 1519
PREDNISOLONE

Methyl Vinyl Ether-Maleic Anhydride Copolymer

1119
EVALUATION OF POLYMERIC MATERIALS. II. SCREENING OF SELECTED VINYLS
AND ACRYLATES AS PROLONGED-ACTION COATINGS
Nessel, R.J. et al.
J. Pharm. Sci., 53, 1964, p. 790
AMPHETAMINE SULPHATE

1120
EVALUATION OF POLYMERIC MATERIALS. III. IN VITRO AND IN VIVO TESTING
OF GRANULES COATED WITH THE n-BUTYL HALF ESTER OF POLYMETHYL VINYL
ETHER/MALEIC ANHYDRIDE
Nessel, R.J. et al.
J. Pharm. Sci., 53, 1964, p. 882
AMPHETAMINE SULPHATE

1121
POLYMERIC PHARMACEUTICAL COATING MATERIALS. I. PREPARATION AND
PROPERTIES
Lappas, L.C. & McKeehan, W.
J. Pharm. Sci., 54, 1965, p. 176

1122
INTERFERON-STIMULATING AND IN VIVO ANTIVIRAL EFFECTS OF VARIOUS
SYNTHETIC ANIONIC POLYMERS
Merigan, T.C. & Finkelstein, M.S.
Virology, 35, 1968, p. 363

1123
MONOMOLECULAR FILM PROPERTIES OF PROTECTIVE AND ENTERIC FILM FOR-
MERS. II. EVAPORATION RESISTANCE AND INTERACTIONS WITH PLASTICISERS
OF POLYMETHYLVINYL ETHER/MALEIC ANHYDRIDE
Zatz, J.L. et al.
J. Pharm. Sci., 58, 1969, p. 1493

1124
CONTROLLED DRUG RELEASE THROUGH POLYMERIC FILMS
Fites, A.L. et al.
J. Pharm. Sci., 59, 1970, p. 610
CAFFEINE; NICOTINIC ACID

1125
DISSOLUTION OF ALKYL VINYLETHER-MALEIC ANHYDRIDE COPOLYMERS AND
ESTER DERIVATIVES
Woodruff, C.W. et al.
J. Pharm. Sci., 61, 1972, p. 1916

1126
EFFECT OF SURFACE ROUGHNESS AND COATING SOLVENT ON FILM
ADHESION TO TABLETS
Nadkarni, P.K. et al.
J. Pharm. Sci., 64, 1975, p. 1554

1127
RIGID SUPPORTS FOR THE IMMOBILISATION OF ENZYMES
Roger, G.P. et al.
J. Macromol. Sci. Chem., A10, 1976, p. 245
TRYPSIN

1128
CONTROLLED DRUG RELEASE BY POLYMER DISSOLUTION. I. PARTIAL ESTERS OF
MALEIC ANHYDRIDE COPOLYMERS — PROPERTIES AND THEORY
Heller, J. et al.
J. Appl. Polym. Sci., 22, 1978, p. 1991
HYDROCORTISONE

VINYLIDENE CHLORIDE-ACRYLONITRILE COPOLYMER

1129
SOLID DOSAGE FORM WITH CONSTANT RELEASE
Lerk, C.F. et al.
J. Pharm. Pharmacol., 25, 1973, (Suppl.), p. 151P
POTASSIUM CHLORIDE; ORPHENADRINE

VINYL FORMAL POLYMER

1130
CONTROLLED SLOW RELEASE OF CHEMOTHERAPEUTIC DRUGS FOR CANCER
FROM MATRICES PREPARED BY RADIATION POLYMERISATION AT LOW
TEMPERATURES
Kaetsu, I.
J. Biomed. Mat. Res., 14, 1980, p. 185
MITOMYCIN; 5-FLUOROURACIL; BLEOMYCIN

VINYLPYRIDINE POLYMERS AND COPOLYMERS

Vinylpyridine Polymer

1131
EFFECTS OF A SERIES OF NEW SYNTHETIC HIGH POLYMERS ON CANCER
METASTASES
Ferruti, P. et al.
J. Med. Chem., 16, 1973, p. 496

1132
POLYVINYLPYRIDINE AND EXPERIMENTAL SILICOSIS
Schlipkoter, H.W. & Brockhaus, A.
Germ. Med. Monthly, 5, 1960, p. 270

1133
SYNTHESIS AND REACTIONS OF HYDROPHILIC FUNCTIONAL MICROSPHERES
FOR IMMUNOLOGICAL STUDIES
Rembaum, A. et al.
J. Macromol. Sci. Chem., A13, 1979, p. 603

Vinylpyridine-Oxide Polymer

1134
INTERACTION OF POLY-2-VINYLPYRIDINE 1-OXIDE AND POLY-4-VINYLPYRIDINE
1-OXIDE WITH MONOSILICIC ACID
Holt, P.F. & Nasrallah, E.T.
J. Chem. Soc., B, 1968, p. 233

1135
POLY(ALKYLVINYLPYRIDINE 1-OXIDES) AND CORRESPONDING ALKYLPYRIDINE
OXIDES: ELECTRONIC SPECTRA AND INTERACTION WITH SILICIC ACID
Holt, P.F. & Lindsay, H.
J. Chem. Soc., B, 1969, p. 54

1136
POLYVINYLPYRIDINE OXIDES IN PNEUMOCONIOSIS RESEARCH
Holt, P.F.
Brit. J. Industr. Med., 28, 1971, p. 72

Vinylpyridine-Vinyl Pyrrolidone Copolymer

1137
UTILISATION OF A MODEL COPOLYMER TO EVALUATE THE CONTRIBUTION OF
HYDROPHOBIC BONDING IN DRUG BINDING
Nagwekar, J.B. & Kostenbauder, H.B.
J. Pharm. Sci., 59, 1970, p. 751
TOLUENE SULPHONIC ACID

VINYL PYRROLIDONE POLYMERS AND COPOLYMERS

Vinyl Pyrrolidone Polymer

1138
PATHOLOGICAL CHANGES IN RABBITS FROM REPEATED INJECTIONS OF PVP
OR DEXTRAN
Nelson, A.H. & Lusky, L.M.
Proc. Soc. Exp. Biol., 76, 1951, p. 765

1139
SPECTRAL STUDIES OF PVP
Oster, G.
J. Polym. Sci., 9, 1952, p. 535
IODINE

1140
CLINICAL STATUS OF DEXTRAN, PVP AND GELATIN PRODUCTS
Pulaski, E.
Quart. Rev. Med., 9, 1952, p. 44

1141
POLYVINYLPYRROLIDONE AS A PLASMA EXPANDER
Ravin, H.A. et al.
N. Engl. J. Med., 247, 1952, p. 921

1142
EFFECTS OF DEXTRAN AND PVP ADMINISTRATION ON LIVER FUNCTION IN
MAN
Reinhold, J.G. et al.
Arch. Surg., 65, 1952, . 706

1143
FATE OF INTRAVENOUSLY ADMINISTERED POLYVINYLPYRROLIDONE
Steele, R. et al.
Ann. N.Y. Acad. Sci., 55, 1952, p. 479

1144
PVP IN SEVERE BURN SHOCK
Cordice, J.W.V. et al.
J. Surg. Gyn. Obstet., 97, 1953, p. 39

1145
LIVER LESIONS FOLLOWING INTRAVENOUS ADMINISTRATION OF POLYVINYLPYRROLIDONE
Gale, E.A. et al.
Amer. J. Clin. Path., 23, 1953, p. 1187

1146
LIVER LESIONS FOLLOWING INTRAVENOUS ADMINISTRATION OF PVP. PHYSIOLOGICAL AND PATHOLOGICAL EFFECTS OF LONG TERM PVP RETENTION
Gall, E.A. et al.
Am. J. Clin. Path., 23, 1953, p. 1187

1147
EXCRETION AND DISTRIBUTION OF POLYVINYLPYRROLIDONE IN MAN
Loeffler, R.K., & Scudder, J.
Am. J. Clin. Pathol., 23, 1953, p. 311

1148
HEXAMETHONIUM IN PVP
Murphy, E.A. & Eastwood, J.
Lancet, 265, 1953, p. 804

1149
PRENDEROL. A NEW DRUG USED WITH RETARDERS IN PETIT MAL EPILEPSY
Perlstein, M.A.
Neurology, 3, 1953, p. 744
2,2-DIETHYL-1,3-PROPANEDIOL

1150
SALPIX: A NEW APPROACH TO THE IDEAL RADIOPAQUE MEDIUM FOR HYSTEROSALPINOGRAPHY
Rubin, I.C. et al.
Fertil. & Steril., 4, 1953, p. 357
SODIUM 3-ACETYLAMINO-2,4,6-TRIIODOBENZOATE; SUSPENSION

1151
INTRAVENOUS USE OF A COMBINATION OF SODIUM p-AMINOSALICYLATE AND PVP IN THE CHEMOTHERAPY OF TUBERCULOISIS
Weiss, W.
Am. J. Med. Sci., 225, 1953, p. 560

1152
FUNCTIONAL AND ANATOMIC EFFECT OF PVP. AN ANALYSIS BASED ON THE STUDY OF 129 CASES
Bernhard, W.G. et al.
Ann. Surg., 139, 1954, p. 397

1153
NECROSIS IN RECTAL CARCINOMA FOLLOWING INTRAVENOUS CHEMOTHERAPHY
Bodkin, L.G.
Am. J. Surg., 88, 1954, p. 367
INDIGOCARMINE; OXALIC ACID; SUSPENSION

1154
ANTIBIOTIC FORMULATIONS
Buckwalter, F.H.
J. Am. Pharm. Assoc., 15, 1954, p. 694
SUSPENSION MEDIUM; PENICILLIN

1155
PVP AS A DRUG RETARDANT. I. EFFECT ON BARBITURATES
Graham, W.D. et al.
J. Pharm. Pharmacol., 6, 1954, p. 27
QUINOLBARBITONE; HEXABARBITONE; PENTOBARBITONE; THIOPENTONE

1156
PVP AS A DRUG RETARDANT. II. EFFECT ON ANALGESICS
Graham, W.D. et al.
J. Pharm. Pharmacol., 6, 1954, p. 115
MORPHINE; PETHIDINE; METHADONE

1157
PVP AS A DRUG RETARDANT. III. EFFECT ON SODIUM p-AMINOSALICYLATE
Graham, W.D. & Teed, H.
J. Pharm. Pharmacol., 6, 1954, p. 558

1158
STUDY OF POSSIBLE COMPLEX FORMATION BETWEEN MACROMOLECULES AND CERTAIN PHARMACEUTICALS
Higuchi, T. & Kuramoto, R.
J. Am. Pharm. Assn., 43, 1954, p. 393
SULPHATHIAZOLE; PROCAIN HCl; SODIUM SALICYLATE; BENZYL PENCILLIN; CHLORAMPHENICOL; MANDELIC ACID; CAFFEINE; THEOPHYLLINE; CORTISONE

1159
STUDY OF POSSIBLE COMPLEX FORMATION BETWEEN MACROMOLECULES AND CERTAIN PHARMACEUTICALS
Higuchi, T. & Kuramoto, R.
J. Am. Pharm. Assn., 43, 1954, p. 398
p-AMINOBENZOIC ACID; AMINOPYRINE; BENZOIC ACID; SALICYLIC ACID; p-HYDROXYBENZOIC ACID; PHENOBARBITAL

1160
STUDY OF POSSIBLE FORMATION BETWEEN MACROMOLECULES AND CERTAIN PHARMACEUTICALS
Guttman, D.E. & Higuchi, T.
J. Amer. Pharm. Assoc. Sci. Ed., 44, 1955, p. 668
IODINE

1161
PVP-IODINE. A THREE YEAR OBSERVATION OF A NEW TOPICAL GERMICIDE
Bogash, R.C.
Bull. Am. Soc. Hosp. Pharm., 13, 1956, p. 226

1162
POSSIBLE COMPLEX FORMATION BETWEEN MACROMOLECULES AND CERTAIN PHARMACEUTICALS. X
Guttman, D.E. & Higuchi, T.
J. Am. Pharm. Assn., 45, 1956, p. 659
PHENOL; RESORCINOL; CATECHOL; HYDROQUINONE; TANNIC ACID; PYROGALLOL; BENZOIC ACID; HYDROXYBENZOIC ACID; SALICYLIC ACID; SODIUM SALICYLATE; METHYL PARABEN; BETA NAPHTHOL; PICRIC ACID

1163
EFFECT OF PVP ON INCIDENCE OF SPONTANEOUS MOUSE MAMMARY CANCER
Stern, K. et al.
Proc. Am. Assoc. Cancer Res., 2, 1956

1164
PHARMACOLOGY OF THE RETICULO-ENDOTHELIAL SYSTEM. I. BLOCKADE BY PVP AS MEASURED WITH RADIO-CHROMIC PHOSPHATE IN THE RABBIT
Weikel, J.H. & Lusky, L.M.
J. Pharmacol. Exp. Therap., 118, 1956, p. 148

1165
EXPERIMENTAL CARCINOGENIC STUDIES IN MACROMOLECULAR CHEMICALS. I. NEOPLASTIC REACTIONS IN RATS AND MICE AFTER PARENTERAL INTRODUCTION OF PVP
Hueper, W.C.
Cancer, 10, 1957, p. 8

1166
FIBROSARCOMAS INDUCED BY MULTIPLE SUBCUTANEOUS INJECTIONS OF CAR-BOXYMETHYL CELLULOSE, POLYVINYLPYRROLIDONE, AND POLYOXYETHYLENE SORBITAN MONOSTEARATE
Lusky, L.M. et al.
Fed. Proc., 16, 1957, p. 318

1167
PVP, A CARCINOGENIC AGENT FOR RATS
Hueper, W.C.
Proc. Am. Assoc. Cancer Res., 2, 1956, p. 120

1168
ANTICOAGULANT EFFECT OF CHLORAZOL FAST PINK AND ITS ANTAGONISM
BY PVP
Burnstein, M. & Laurent, A.
Arch. Intern. Pharmacodyn., 112, 1957, p. 395

1169
POVIDONE-IODINE AS A TOPICAL ANTISEPTIC
Gershenfeld, L.
Am. J. Surg., 94, 1957, p. 938

1170
CHEMISTRY OF PVP-IODINE
Siggia, S.
J. Am. Pharm. Assoc., 46, 1957, p. 201

1171
LYMPH NODE CHANGES DUE TO PVP
Towers, R.P.
J. Clin. Path. (London), 10, 1957, p. 175
HEXAMETHONIUM BROMIDE

1172
POVIDONE-IODINE
J. Am. Med. Assoc., 163, 1957, p. 647

1173
STUDY OF TABLET COATING USING POLYVINYLPYRROLIDONE AND
ACETYLATED MONOGLYCERIDE
Ahsan, S.S. & Blaug, S.M.
Drug Standards, 26, 1958, p. 29

1174
EFFECTS OF VARIOUS SKIN-STERILISATION TECHNIQUES ON THE VIABILITY OF
SKIN
Georgiade, N. et al.
Plast. & Reconstruct. Surg., 21, 1958, p. 479
IODINE

1175
EVALUATION OF POST MORTEM SURVIVAL OF SKIN BY TISSUE CULTURE
METHODS
Kepes, J.
Plast. & Reconstruct. Surg., 21, 1958, p. 483
IODINE

1176
CLINICAL EVALUATION OF POVIDONE-IODINE AEROSOL SPRAY IN SURGICAL
PRACTICE
Garnes, A.L. et al.
Am. J. Surg., 97, 1959, p. 49

1177
INTERACTION OF PRESERVATIVES WITH MACROMOLECULES. III.
PARAHYDROXYBENZOIC ACID ESTERS IN THE PRESENCE OF SOME HYDROPHILIC
POLYMERS
Miyawaki, G.M. et al.
J. Am. Pharm. Assn., 48, 1959, p. 315
PARABENS

1178
FECAL PVP EXCRETION IN HYPOALBUMINAEMIA AND GASTROINTESTINAL
DISEASE
Dawson, A.M. et al.
Brit. Med. J., 5253, 1961, p. 667

1179
EFFECT OF POVIDONE-IODINE ON CUTANEOUS INFECTIONS AND THE FLORA
OF THE SKIN
Frank, L.
Therap. Res., 3, 1961, p. 535

1180
COMPARATIVE LABORATORY EVALUATION OF ANTISEBORRHEIC DERMATITIS
PREPARATION
Greenberg, L.
J. Pharm. Sci., 50, 1961, p. 480
IODINE

1181
BIOASSAY OF POLYVINYLPYRROLIDONE WITH LIMITED MOLECULAR WEIGHT
RANGE
Hueper, W.C.
J. Nat. Cancer Inst., 26, 1961, p. 229

1182
IMPROVED SMALLPOX VACCINE
Amies, C.R.
J. Hyg. Camb., 60, 1962, p. 473
SUSPENSION MEDIUM

1183
REVIEW OF THE PHYSIOLOGICAL PROPERTIES OF POLYVINYLPYRROLIDONE
Burnette, L.W.
Proc. Sci. Sect. Toilet Goods Assoc., 38, 1962, p. 1

1184
FACIAL BURNS
Georgiade, N.G. et al.
Plast. Reconstruct. Surg., 29, 1962, p. 648
IODINE

1185
POVIDONE-IODINE AS A SPORICIDE
Gershenfeld, L.
Am. J. Pharmacy, 134, 1962, p. 78

1186
POVIDONE-IODINE AS A VAGINAL MICROBICIDE
Gershenfeld, L.
Am. J. Pharmacy, 134, 1962, p. 278

1187
POVIDONE-IODINE AS A TRICHOMONACIDE
Gershenfeld, L.
Am. J. Pharmacy, 134, 1962, p. 324

1188
STUDY OF DISINFECTION OF THE SKIN. A COMPARISON OF POVIDONE-
IODINE WITH OTHER AGENTS USED FOR SURGICAL SCRUBS
Joress, S.M.
Ann. Surg., 155, 1962, p. 296

1189
ANTIBACTERIAL AGENTS NOT PRESENTLY EMPLOYED AS PRESERVATIVES IN
OPHTHALMIC PREPARATIONS FOUND EFFECTIVE AGAINST PSEUDOMONAS
AERUGINOSA
Kohn, S.R.
J. Pharm. Sci., 52, 1963, p. 1126
IODINE

1190
ACTIVITY OF GLUCOSE-6-PHOSPHATASE, ADENOSINE TRIPHOSPHATASE, SUC-
CINIC DEHYDROGENASE, AND ACID PHOSPHATASE AFTER DEXTRAN OR PVP
UPTAKE BY LIVER IN VIVO
Meijer, A.E. & Willihagen, R.G.
Biochem. Pharmacol., 12, 1963, p. 973

1191
EFFECT OF TOPICAL POVIDONE-IODINE (BETADINE) ON SERUM PROTEIN-
BOUND IODINE
Quagliana, J.M.
J. Clin. Endocrinol. Metab., 23, 1963, p. 395

1192
CLINICAL STUDY OF A POVIDONE-IODINE REGIMEN FOR RESISTANT VAGINITIS
Shook, D.M.
Curr. Therap. Res., 5, 1963, p. 256

1193
PREOPERATIVE SKIN PREPARATION WITH POVIDONE-IODINE
Close, A.S. et al.
Am. J. Surg., 108, 1964, p. 398

1194
POVIDONE-IODINE. EXTENSIVE SURGICAL EVALUATION OF A NEW ANTISEPTIC
AGENT
Connell, J.F. & Rousselot, L.M.
Am. J. Surg., 108, 1964, p. 849

1195
EFFECT OF POVIDONE-IODINE (BETADINE) ON SERUM PROTEIN-BOUND
IODINE, WHEN USED AS A SURGICAL PREPARATION ON INTACT SKIN
Higgins, H.P. et al.
Canadian Med. Assn. J., 90, 6 June 1964, p. 1298

1196
METHODS FOR DISINFECTION OF HANDS AND OPERATION SITES
Lowbury, E.J. et al.
Brit. Med. J., 2, 1964, p. 531
IODINE

1197
EVALUATION OF POLYMERIC MATERIALS. III. IN VITRO AND IN VIVO TESTING
OF GRANULES COATED WITH THE n-BUTYL HALF ESTER OF POLYMETHYL VINYL
ETHER/MALEIC ANHYDRIDE
Nessel, R.J. et al.
J. Pharm. Sci., 53, 1964, p. 882
AMPHETAMINE SULPHATE

1198
PVP-IODINE. AN ASSESSMENT OF ANTIBACTERIAL ACTIVITY
Saggars, B.A. & Stewart, G.R.
J. Hyg., 62, 1964, p. 508

1199
PVP FOR TABLET MAKING
Prescott, F.J.
Drug Cos. Ind., 97, 1965, p. 497

1200
TRIAL OF POVIDONE-IODINE OINTMENT IN THE TREATMENT OF LEG ULCERS
Thorne, N. & Fox, D.
Practitioner, 194, 1965, p. 250

1201
EVALUATION OF POLYMERIC MATERIALS. IV. GRANULATING AGENTS FOR
COMPRESSED TABLETS
Willis, C.R. et al.
J. Pharm. Sci., 54, 1965, p. 366

1202
EFFECTS OF POVIDONE-IODINE IN THE TREATMENT OF BURNS AND
TRAUMATIC LOSSES OF SKIN
Wynn-Williams, D. & Monballiu, G.
Brit. J. Plast. Surg., 18, 1965, p. 146

1203
EVALUATION OF POLYMERIC MATERIALS. IV. GRANULATING AGENTS FOR
COMPRESSED TABLETS
J. Pharm. Sci., 54, 1965, p. 366

1204
INTERACTION OF PVP WITH IODINE
Eliassaf, J.
Eur. Polym. J., 2, 1966, p. 269

1205
COMPARISON OF THE HYPOLIPIDEMIC ACTION OF DEXTRAN AND POLYVINYL-
PYRROLIDONE
Sanbar, S.S. & Smet, G.
J. Lab. Clin. Med., 70, 1967, p. 890

1206
INTERACTION OF SODIUM ERYTHROSIN AND POLYVINYLPYRROLIDONE
Anderson, J.C. & Boyce, G.M.
J. Pharm. Sci., 58, 1969, p. 1425

1207
DISSOLUTION CHARACTERISTICS OF RESERPINE-POLYVINYLPYRROLIDONE
COPRECIPITATES
Bates, T.R.
J. Pharm. Pharmacol., 21, 1969, p. 710

1208
BLOOD REPLACEMENT
Gruber, U.F.
Springer;Verlag, 1969
PLASMA EXPANDER

1209
COMPLEXATION BETWEEN HEXYLRESORCINOL AND POLYVINYLPYRROLIDONE
AND ITS MICROBIAL EVALUATION
Polli, G.P. & Frost, B.M.
J. Pharm. Sci., 58, 1969, p. 1543

1210
DISSOLUTION RATES OF HIGH ENERGY POLYVINYLPYRROLIDONE-
SULPHATHIAZOLE COPRECIPITATES
Simonelli, A.P. et al.
J. Pharm. Sci., 58, 1969, p. 538
SULPHATHIAZOLE

1211
INHIBITION OF SULPHATHIAZOLE CRYSTAL GROWTH BY
POLYVINYLPYRROLIDONE
Simonelli, A.P. et al.
J. Pharm. Sci., 59, 1970, p. 633

1212
EVIDENCE FOR THYMUS INDEPENDENT HUMORAL ANTIBODY PRODUCTION IN
MICE AGAINST POLYVINYLPYRROLIDONE AND E. COLI LIPOPOLYSACCHARIDE
Andersson, B. & Blomgren, H.
Cell. Immunol., 2, 1971, p. 411

1213
STUDIES ON THE MOLECULAR WEIGHT AND CRYOPROTECTIVE PROPERTIES OF
POLYVINYLPYRROLIDONE AND DEXTRAN WITH BACTERIA AND ERYTHROCYTES
Ashwood-Smith, M.J. & Warby, C.
Cryobiology, 8, 1971, p. 453

1214
INTERACTION OF BISHYDROXYCOUMARIN WITH POLYVINYL PYRROLIDONE
Cho, M.J. et al.
J. Pharm. Sci., 60, 1971, p. 720
BISHYDROXYCOUMARIN

1215
SIDE EFFECTS OF POTASSIUM CHLORIDE IN PRODUCTS WITH DIFFERENT
DISSOLUTION RATES
Graffner, C. & Sjogren, J.
Acta Pharm. Suec., 8, 1971, p. 19

1216
ANTITUMOUR ACTIVITY OF AN ACRONYCINE-POLYVINYLPYRROLIDONE
COPRECIPITATE
Svoboda, G.H. et al.
J. Pharm. Sci., 60, 1971, p. 333

1217
MONOLAYERS OF POLYVINYLPYRROLIDONE COPOLYMERS
Zatz, J.L. & Knowles, B.
J. Pharm. Sci., 60, 1971, p. 1731

1218
EFFECT OF ADJUVANTS ON TACKINESS OF POLYVINYLPYRROLIDONE FILM
COATING
Alan, A.S. et al.
J. Pharm. Sci., 61, 1972, p. 265

1219
MODEL REACTIONS FOR SYNTHESIS OF PHARMACOLOGICALLY ACTIVE
POLYMERS BY WAY OF MONOMERIC AND POLYMERIC REACTIVE ESTERS
Batz, H.G. et al.
Angew. Chem. Int. Edn., 11, 1972, p. 1103

1220
ENHANCED ABSORPTION AND DISSOLUTION OF RESERPINE FROM RESERPINE-
POLYVINYLPYRROLIDONE COPRECIPITATES
Stupak, E.I. & Bates, T.R.
J. Pharm. Sci., 61, 1972, p. 400

1221
POLYMERS CONTAINING PHENETHYLAMINES
Weiner, B.Z. et al.
J. Med. Chem., 15, 1972, p. 410

1222
PENETRATION OF POLYMER MONOLAYERS
Zatz, J.L.
J. Pharm. Sci., 61, 1972, p. 977

1223
NEW TABLET DISINTEGRATING AGENT: CROSS-LINKED POLYVINYL-
PYRROLIDONE
Kornblum, S.S. & Stoopak, S.B.
J. Pharm. Sci., 62, 1973, p. 43

1224
STUDIES ON MICROENCAPSULATED POWDERS
Nath, B.S. et al.
Ind. J. Pharmacy, 35, 1973, p. 131
MAGNESIUM CARBONATE

1225
POLYVINYLPYRROLIDONE AS A SOLUBLE CARRIER OF PROTEINS
Specht, B. et al.
Physiol. Chem., 354, 73, p. 1659
TRYPSIN

1226
DIFFUSION IN POLYMER GEL IMPLANTS
Davis, B.K.
Proc. Nat. Acad. Sci., USA, 71, 1974, p. 3120
BOVINE SERUM ALBUMIN; IMMUNOGLOBULIN; INSULIN; LUTENISING HOR-
MONE; PROSTAGLANDIN F2-ALPHA

1227
ENZYME TRAPPING IN PVP: PRELIMINARY REPORT
Denti, E.
Biomat. Med. Dev. Artif. Org., 2, 1974, p. 293
ASPARAGINASE

1228
MANAGEMENT OF ACUTE GLAUCOMA WITH PILOCARPINE SOAKED
HYDROPHILIC LENS
Hillman, J.S.
Brit. J. Ophthalmol., 58, 1974, p. 674

1229
GEL ENTRAPMENT OF ENZYMES: KINETIC STUDIES OF IMMOBILISED GLUCOSE
OXIDASE
Hinberg, I. et al.
Biotech. Bioeng., 16, 1974, p. 159

1230
INTRAVAGINAL INSERTION OF A DIMETHYLPOLYSILOXANE-POLYVINYL-
PYRROLIDIONE-PROSTAGLANDIN F2 ALPHA TUBE FOR MID TERM ABORTION IN
RABBITS
Lau, I.F. et al.
Am. J. Obstet. Gynecol., 120, 1974, p. 837

1231
MID TERM ABORTION WITH SILASTIC-PVP IMPLANT CONTAINING
PROSTAGLANDIN F2 ALPHA IN RABBITS, RATS AND HAMSTERS
Lau, I.F. et al.
Fertil. & Steril., 25, 1974, p. 839

1232
SUPPRESSION OF BENZOIC ACID ADSORPTION ON SULPHAMETHAZINE BY
POLYVINYLPYRROLIDONE: EFFECT OF CONTACT TIME
Nasipuri, R.N. & Khalil, S.A.H.
J. Pharm. Sci., 63, 1974, p. 1788

1233
PROSTAGLANDIN E2: MID TERM ABORTION IN RATS USING SILASTIC-PVP TUBES
Saksena, S.K. et al.
Prostaglandins, 7, 1974, p. 507

1234
PROSTAGLANDIN F2 ALPHA IMPLANT-INDUCED ABORTION: EFFECT OF
PROGESTIN AND LUTENISING HORMONE CONCENTRATION AND ITS REVERSAL
BY PROGESTERONE IN RABBITS, RATS AND HAMSTERS
Saksena, S.V. et al.
Fertil. & Steril., 25, 1974, p. 845

1235
INFLUENCE OF POLYVINYLPYRROLIDONE ON THE DISSOLUTION PROPERTIES
OF HYDROFLUMETHIAZIDE
Corrigan, O.I. & Timoney, R.F.
J. Pharm. Pharmacol., 27, 1975, p. 759

1236
MACROMOLECULAR DRUGS ACTING AS PRECURSORS OF NON-
MACROMOLECULAR ACTIVE SUBSTANCES — PRELIMINARY CONSIDERATIONS
Ferruti, P.
Pharmacol. Res. Comm., 7, 1975, p. 1

1237
INDUCTION OF TEMPORARY STERILITY IN FEMALE HAMSTERS BY AN
INTRAPERITONEAL SILASTIC-PVP-PG F2 ALPHA TUBE
Lau, I.F. et al.
Prostaglandins, 9, 1975, p. 893

1238
SOLUBLE OPHTHALMIC DRUG INSERTS
Maichuk, Y.F.
Lancet, 1, 18 Jan. 1975, p. 173
NEOMYCIN; KANAMYCIN; SULPHAMETHOXYPYRIDAZINE; IDOXURIDINE;
FLORENAL; ATROPINE; PILOCARPINE; DEXAMETHASONE

1239
APPLICATION OF POLYVINYLPYRROLIDONE AS A CARRIER FOR KALLIKREIN
Specht, B.U. et al.
Arch. Int. Pharmacodyn., 213, 1975, p. 242

1240
GENETIC EFFECTS OF POVIDONE-IODINE
Wlodkowski, T.J. et al.
J. Pharm. Sci., 64, 1975, p. 1235

1241
DEVELOPMENT OF A STABLE SUBLINGUAL NITROGLYCERIN TABLET II:
FORMULATION AND EVALUATION OF TABLETS CONTAINING POVIDONE
Fung, H.L. et al.
J. Pharm. Sci., 65, 1976, p. 558

1242
RELEASE OF INORGANIC FLUORIDE ION FROM RIGID POLYMER MATRICES
Halpern, B.D. et al.
ACS Coatings & Plastics Preprints, 36, 1976, p. 341
c.f. ACS Symp. Series, 33, 1976, p. 135

1243
SUSTAINED RELEASE OF MACROMOLECULES FROM POLYMERS
Langer, R. & Folkman, J.
Midland Macromol. Inst., Polymeric Delivery Systems, 5th Int. Symp., Aug.
1976, p. 175
SOYBEAN TRYPSIN INHIBITOR; LYSOZYME; ALKALINE PHOSPHATASE;
CATALASE; INSULIN; HEPARIN

1244
POLYMERS FOR THE SUSTAINED RELEASE OF PROTEINS AND OTHER
MACROMOLECULES
Langer, R. & Folkman, J.
Nature, 263, Oct. 28, 1976, p. 797

1245
POLYVINYLPYRROLIDONE-STORAGE DISEASES
Reske-Nelson, E. et al.
Acta Path. Microbiol. Scand., 84A, 1976, p. 397

1246
DISSOLUTION RATES OF HIGH ENERGY SULPHATHIAZOLE-POVIDONE
COPRECIPITATES. II. CHARACTERISATION OF FORM OF DRUG CONTROLLING ITS
DISSOLUTION RATE VIA SOLUBILITY STUDIES
Simonelli, A.P. et al.
J. Pharm. Sci., 65, 1976, p. 355

1247
FLUORESCENT PROBE STUDY OF SULPHONAMIDE BINDING TO POVIDONE
Hsiao, C.H. et al.
J. Pharm. Sci., 66, 1977, p. 1157

1248
SYNTHESIS OF POLYMERS WITH COVALENTLY BOUND DRUGS
Kropachev, V.A.
ACS Polymer Preprints, 18, No. 1, 1977, p. 557
ISONICOTINIC ACID; AMPICILLIN; ACETYLSALICYLIC ACID; 4-OXYCUMARINE;
NOVOCAINE; DICAINE

1249
STUDY OF POLYMER DRUG INTERACTED SYSTEMS AS A PHYSICO-CHEMICAL AP-
PROACH IN DRUG DESIGN. I. A STUDY OF DIPHENHYDRAMINE ENTRAPMENT
COMPOSITIONS AS A NEW CONTRIBUTION TO CONTROLLED RELEASE ORAL
DRUGS
Salib, N.N.
1st Int. Conf. Pharm. Technol., 1977, p. 75

1250
CO-PRECIPITATED SYSTEMS OF SALICYL SALICYLIC ACID WITH POLYVINYL-
PYRROLIDONE: PHYSICAL AND BIOAVAILABILITY STUDIES
Stevens, L.A. & Padfield, J.M.
1st Int. Conf. Pharm. Technol., 1977, p. 135

1251
STORAGE OF POLYVINYLPYRROLIDONE (PVP) IN TISSUES FOLLOWING LONG
TERM TREATMENT WITH A PVP-CONTAINING VASOPRESSIN PREPARATION
Christensen, M. et al.
Acta Med. Scand., 204, 1978, p. 295

1252
SUSTAINED RELEASE FROM INERT WAX MATRICES. III. EFFECT OF POVIDONE
ON TRIPELENNAMINE HYDROCHLORIDE RELEASE
Dakkuri, A. et al.
J. Pharm. Sci., 67, 1978, p. 357

1253
MOLECULAR INTERACTION BETWEEN E-PROSTAGLANDINS AND SELECTED
POLYMERS AND ITS POTENTIAL UTILISATION IN ORAL DOSAGE FORM DESIGN
Fung, H.L. & Cho, M.J.
J. Pharm. Sci., 67, 1978, p. 971

1254
POLYMERIC AFFINITY DRUGS FOR CARDIOVASCULAR, CANCER AND
UROLITHIASIS THERAPY
Goldberg, E.P.
Polymeric Drugs, 1978, p. 239, Academic Press
Donaruma, L.G. & Vogl, O. eds.

1255
SYNTHETIC POLYMERS IN CHEMOTHERAPY: GENERAL PROBLEMS
Kalal, J. et al.
Polymeric Drugs, 1978, p. 131, Academic Press
Donaruma, L.G. & Vogl, O. eds.

1256
POLYMERS AS INTERFERON INDUCERS
Levy, H.B.
Polymeric Drugs, 1978, p. 305, Academic Press
Donaruma, L.G. & Vogl, O. eds.

1257
SYNTHESIS AND SOME PROPERTIES OF ANTITHROMBOGENIC POLYMERS
Plate, N.A.
Polymeric Drugs, 1978, p. 63, Academic Press
Donaruma, L.G. & Vogl, O. eds.

1258
POLYMERIC DRUGS IN THE CHEMOTHERAPY OF MICROBIAL INFECTIONS
Samour, C.M.
Polymeric Drugs, 1978, p. 161, Academic Press
Donaruma, L.G. & Vogl, O. eds.

1259
NITROGLYCERIN STABILITY IN POLYETHYLENE GLYCOL 400 AND POVIDONE
SOLUTIONS
Suphajettra, P. et al.
J. Pharm. Sci., 67, 1978, p. 1394
NITROGLYCERIN

1260
CONTROLLED CHEMOTHERAPY THROUGH MACROMOLECULES
Zaffaroni, A. & Bonson, P.
Polymeric Drugs, 1978, p. 1, Academic Press
Donaruma, L.G. & Vogl, O. eds.

1261
CONTROLLED RELEASE OF DRUGS FROM HYDROGEL MATRICES
Hosaka, S. et al.
J. Appl. Polym. Sci., 23, 1979, p. 2089
ERYTHROMYCIN; ERYTHROMYCIN ESTOLATE

1262
POSSIBILITIES OF OBTAINING POLYMERS WITH SUPPOSED ANTITUMOUR
ACTIVITY
Iliev, I.B. et al.
J. Polym. Sci. Polym. Symp., No. 66, 1979, p. 1

1263
MODIFICATION OF FIBRIN BY SYNTHETIC POLYMERS
Kovacs, G. et al.
J. Polym. Sci. Polym. Symp., No. 66, 1979, p. 201

1264
TERMINATION OF PSEUDOPREGNANCY IN RATS BY SILASTIC-PVP-PG F2 ALPHA
TUBE
Lau, I.F.
Prostaglandins Med., 2, 1979, p. 373

1265
ABSENCE OF POVIDONE-IODINE-INDUCED MUTAGENICITY IN MICE AND HAMSTERS
Merkle, J. & Zeller, H.
J. Pharm. Sci., 68, 1979, p. 100

1266
SYNTHESIS AND REACTIONS OF HYDROPHILIC FUNCTIONAL MICROSPHERES FOR IMMUNOLOGICAL STUDIES
Rembaum, A. et al.
J. Macromol. Sci. Chem., A13, 1979, p. 603

1267
INCREASED SERUM IODIDE CONCENTRATION FROM IODINE ABSORPTION THROUGH WOUNDS TREATED TOPICALLY WITH POVIDONE-IODINE
Aronoff, G.R. et al.
Am. J. Med. Sci., 279, 1980, p. 173

1268
DISSOLUTION STUDIES OF POVIDONE-SULPHATHIAZOLE COACERVATED SYSTEMS
Badawi, A.A. & El-Sayed, A.A.
J. Pharm. Sci., 69, 1980, p. 492

1269
ULTRASTRUCTURAL EFFECTS OF THE HIGH MOLECULAR WEIGHT CRYOPROTEC-TANTS DEXTRAN AND POLYVINYLPYRROLIDONE ON LIVER AND BROWN ADIPOSE TISSUE IN VITRO
Barnard, T.
J. Microsc., 120, 1980, p. 93

1270
NEW GROUP OF SYNTHETIC POLYMERS WITH ANTITUMOUR ACTIVITY
Bierling, R. et al.
Naturwissenschaften, 67, 1980, p. 366

1271
SEPARATION OF HUMAN EOSINOPHILS IN DENSITY GRADIENTS OF POLYVINYLPYRROLIDONE-COATED SILICA GEL (PERCOLL)
Gartner, I.
Immunology, 40, 1980, p. 133

1272
ISOLATION OF A STABLE CELL WALL-DEFECTIVE FORM OF NEISSERIA GONORRHOEAE FROM A CASE OF UNTREATED GONOCOCOAL URETHRITIS
Hickman, R.K. & Lawson, J.W.
J. Clin. Microbiol., 12, 1980, p. 603

1273
MODEL OF INTERACTION OF AJMALINE WITH POLYVINYLPYRROLIDONE
Hosono, T. et al.
J. Pharm. Sci., 69, 1980, p. 824
COPRECIPITATE

1274
IMMOBILISATION OF ENZYMES WITH USE OF PHOTOSENSITIVE POLYMERS
HAVING THE STILBAZOLIUM GROUP
Ichimura, K. & Watanabe, S.
J. Polym. Sci. Polym. Chem., 18, 1980, p. 891
INVERTASE; GLUCOAMYLASE; CATALASE

1275
COMPARATIVE TRIAL OF CEPHALORIDINE WITH POLYBACTRIA, POVIDONE-
IODINE AND CONTROL FOR PROPHYLAXIS OF WOUND INFECTION IN
ABDOMINAL SURGERY
Kenning, B.R. et al.
Brit. J. Clin. Pract., 34, 1980, p. 45

1276
INTERACTION OF POVIDONE WITH AROMATIC COMPOUNDS. I. EVALUATION
OF COMPLEX FORMATION BY FACTORIAL ANALYSIS
Plaizier-Vercammen, J.A. & DeNeve, R.E.
J. Pharm. Sci., 69, 1980, p. 1403
SALICYLIC ACID; BENZOIC ACID

1277
POVIDONE-IODINE ANTISEPSIS FOR TRANSRECTAL PROSTATIC BIOPSY
Rees, M. et al.
Br. Med. J., 281, Sept. 6, 1980, p. 650

1278
INCREASED GASTROINTESTINAL ABSORPTION OF LARGE MOLECULES IN
PATIENTS AFTER 5-FLUOROURACIL THERAPY FOR METASTATIC COLON
CARCINOMA
Siber, G.R. et al.
Cancer Res., 40, 1980, p. 3430

1279
REGULATION OF THE IMMUNE RESPONSE TO POLYVINYLPYROLLIDONE: AN-
TIGEN INDUCED CHANGES IN PROSTAGLANDIN AND CYCLIC NUCLEOTIDE
LEVELS
Zimecki, M. et al.
Arch. Immunol. Ther. Exp., 28, 1980, p. 179

Vinylpyrrolidone-Maleic Anhydride Copolymer
and
Vinylpyrrolidone-Methacrylic Acid Copolymer

1280
POSSIBILITIES OF OBTAINING POLYMERS WITH SUPPOSED ANTITUMOUR
ACTIVITY
Iliev, I.B. et al.
J. Polym. Sci. Polym. Symp., No. 66, 1979, p. 1
MONOCHLOROETHYLAMINE

VINYL SULPHATE POLYMER

1281
ANTIVIRAL ACTIVITY OF AN INTERFERON INDUCING SYNTHETIC POLYMER
Carne, P.E. et al.
Proc. Soc. Exp. Biol. & Med., 131, 1969, p. 443

1282
IN VITRO ACTION OF POLYVINYLSULPHATE ON ACYLATION AND
METHYLATION OF NUCLEIC ACIDS
Abeels, M.J.F.
Experientia, 26, 1970, p. 483
RIBONUCLEASE INHIBITOR

1283
POLYMERS AS INTERFERON INDUCERS
Levy, H.B.
Polymeric Drugs, 1978, p. 305, Academic Press
Donaruma, L.G. & Vogl, O. eds.

VINYL SULPHONIC ACID POLYMER

1284
HYDROLYSIS OF PEPTIDES AND PROTEINS WITH POLYVINYLSULPHONIC ACID
Kern, W. & Scherhag, B.
Makromol. Chem., 28, 1958, p. 209

XENYLPHOSPHATE POLYMER

1285
ANIONIC POLYMERS. III. TISSUE DISTRIBUTION AND METABOLISM OF
POLYXENYLPHOSPHATE IN TUMOUR BEARING MICE
Muelbaecher, C. et al.
Cancer Res., 19, 1959, p. 907

1286
ANIONIC POLYMERS. IV. MICROELECTROPHORESIS OF ASCITES TUMOUR CELLS
AND THE EFFECTS OF POLYXENYLPHOSPHATE
Staumfjord, J.V. & Hummel, J.P.
Cancer Res., 19, 1959, p. 913

APPENDIX

Just prior to publication a number of references were brought to my attention which I considered too important to be excluded. They are listed here as an appendix in the same format as the main body of the book. The majority of references were kindly supplied by the ALZA Corporation who will, upon request, supply updated bibliographies concerning their therapeutic systems. Details may be obtained from ALZA Corporation, 950 Page Mill Road, Palo Alto, California 94304, U.S.A.

GENERAL

A1
RECENT DEVELOPMENTS IN DOSAGE FORM DESIGN
Davis, S.S.
Drug Dev. Comm., 2, 1976, p. 151

A2
NEW CONCEPTS AND STANDARDS OF QUALITY CONTROL AS APPLIED TO CONTROLLED DRUG DELIVERY SYSTEMS
Michaels, A.S. et al.
Quality Control of Medicines, Chapter 3, 1976, p. 45
Deasey, P.B. & Timoney, R.F. eds.
Elsevier/N. Holland Biomed. Press
PILOCARPINE; PROGESTERONE

A3
SYNTHETIC POLYMERIC MEMBRANES: PRACTICAL APPLICATIONS — PAST, PRESENT AND FUTURE
Michaels, A.S.
Pure Appl. Chem., 46, 1976, p. 193

A4
REGULATION OF HUMAN FERTILITY
Moghissi, K.S. & Evans, T.N. eds.
Wayne State University Press, Detroit, 1976

A5
DRUG DELIVERY SYSTEMS OF THE FUTURE
Swarbrick, J.
Aust. J. Pharm. Sci., NS5, 1976, p. 73

A6
SYM. PROC.: CLIN. EXP. WITH THE PROGESTERONE UTERINE THERAPEUTIC SYSTEM
ACAPULCO, Oct. 15—16, 1976
Exerpta Medica

A7
MEMBRANE TECHNOLOGY: PRINCIPLES AND THERAPEUTIC POSSIBILITIES OF CONTROLLED DELIVERY OF DRUGS
Heilmann, K.
Documenta Ophthalmol. Proc. Series, 12, 1977, p. 7

A8
THERAPEUTIC SYSTEMS — PATTERN-SPECIFIC DRUG DELIVERY: CONCEPT AND DEVELOPMENT
Heilmann, K., 1978
Georg Thieme, Stuttgart

A9
DRUG DELIVERY SYSTEMS: A BRIEF REVIEW
Juliano, R.L.
Can. J. Physiol. Pharmacol., 56, 1978, p. 683

A10
NEW SYSTEMS FOR DRUG DELIVERY
Shaw, J.E. & Theeuwes, F.
Aust. J. Pharm. Sci., 7, 1978, p. 49

A11
THERAPEUTIC IMPLICATIONS OF CONTROLLED DRUG DELIVERY
Zaffaroni, A.
Future Trends in Therapeutics, 1978, p. 143
McMahon, F.G. (ed.)
Futura Pub. Co., New York

A12
THERAPEUTIC SYSTEMS: THE KEY TO RATIONAL DRUG THERAPY
Zaffaroni, A.
Drug Metab. Rev., 8, 1978, p. 191

A13
CONTROLLED DRUG DELIVERY RATE SYSTEMS
Gale, R.M.
Vortex, 40, 1979, p. 12

A14
TAKING OUR MEDICINE
Ganderton, D.
Pharm. J., 223, 1979, p. 245

A15
CONTROLLED RELEASE OF BIOACTIVE MATERIALS
Baker, R. (ed.)
Academic Press, New York, 1980

A16
MEDICATED INTRAUTERINE DEVICES: PHYSIOLOGICAL AND CLINICAL CONCEPTS
Hafez, E.S.E. & Van Os, W.A.A. eds.
Martinus Nijhoff, The Hague, Netherlands, 1980

A17
PROGRESS IN CONTRACEPTIVE DELIVERY SYSTEMS: IUD PATHOLOGY AND MANAGEMENT
Hafez, E.S.E. & Van Os, W.A.A. eds.
G.K. Hall Med. Pub., Boston, 1980

A18
POLYMERIC DRUG DELIVERY SYSTEMS
Kim, S.W. et al.
Drug Design, 10, 1980
Ariens, E.J. ed., Medicinal Chemistry Series, Academic Press

A19
POLYMERIC DELIVERY SYSTEMS FOR CONTROLLED DRUG RELEASE
Langer, R.
Chem. Eng. Commun., 6, 1980, p. 1

A20
DIFFUSIONAL RELEASE OF A SOLUTE FROM A POLYMERIC MATRIX — APPROXIMATE ANALYTICAL SOLUTIONS
Lee, P.I.
J. Memb. Sci., 7, 1980, p. 255
KINETIC ANALYSIS

A21
DRUG DELIVERY SYSTEMS
Shaw, J.E.
Ann. Reports in Med. Chem., 15, 1980, p. 302
Hess, H.J. (ed.)
Academic Press, New York

A22
NEW PHARMACEUTICAL DOSAGE FORMS
Urquhart, J. & Benson, H.
Lancet, 1 (8164), 1980, p. 367

A23
CONTROLLED RELEASE OF LONG-DURATION DRUGS ENHANCES MEDICAL
BENEFIT
Urquhart, J.
Med. Trib., 21, 1980, p. 7

A24
DEVELOPMENT OF CONTROLLED DELIVERY DRUG SYSTEMS
Zaffaroni, A.
The Chemist, 57, 1980, p. 4

ACRYLAMIDE POLYMER

A25
O-GLYCOSYL POLYACRYLAMIDE GELS: APPLICATION TO PHYTOHEMAGGLUTININS
Horejsi, V. & Kocourek, J.
Methods in Enzymology: Affinity Techniques, 34, 1974, p. 361
Jakoby, W.B. & Wilchek, M. eds.
Academic Press, New York

A26
COVALENT LINKAGE OF FUNCTIONAL GROUPS, LIGANDS, AND PROTEINS TO POLYACRYLAMIDE BEADS
Inman, J.K.
Methods in Enzymology: Affinity Techniques, 34, 1974, p. 30
Jakoby, W.B. & Wilchek, M. eds.
Academic Press, New York

METHYL ACRYLATE POLYMER

A27
POLYMERS COUPLED TO AGAROSE AS STABLE AND HIGH CAPACITY SPACERS
Wilchek, M. & Miron, T.
Methods in Enzymology: Affinity Techniques, 34, 1974, p. 72
Jakoby, W.B. & Wilchek, M. eds.
Academic Press, New York

ETHYLENEIMINE POLYMER

A28
MICROCAPSULES FOR CONTROLLED RELEASE VIA DONNAN DIALYSIS
Cabasso, I.
J. Memb. Sci., 7, 1980, p. 305
SODIUM NITRATE; POTASSIUM BICARBONATE; SODIUM CHLORIDE; SODIUM PHOSPHATE

ETHYLENE-VINYL ACETATE COPOLYMER

A29
SIMULATED SUSTAINED RELEASE PILOCARPINE THERAPY AND AQUEOUS
HUMOR DYNAMICS
Lerman, S. & Reininger, B.
Can. J. Ophthalmol., 6, 1971, p. 14

A30
NEW APPROACH TO DRUG ADMINISTRATION
Place, V.
Development and Control of New Drug Products, 1972, p. 41
Pernarowski, M. & Darrach, M. (eds.)
Evergreen Press Ltd.
PILOCARPINE; PROGESTERONE

A31
RELEASE OF H-PROGESTERONE FROM POLYMERIC SYSTEMS IN THE RHESUS
MONKEY
Kulkarni, B.D. et al.
Contraception, 8, 1973, p. 299

A32
ONE-YEAR EXPERIENCE WITH DIFFERENT RELEASE RATES OF THE UTERINE
PROGESTERONE DELIVERY SYSTEM FOR CONTRACEPTION
Aznar-Ramos, R. et al.
Fertil. Steril., 25, 1974, p. 308

A33
PILOCARPINE OCUSERT DELIVERY SYSTEM
Flach, A.
Trans. Pac. Coast Otoophthalmol. Soc., 55, 1974, p. 179

A34
NEW DRUG DELIVERY SYSTEM
Fraunfelder, F.T.
Ophthalmol. Digest., Nov. 1974, p. 27
PILOCARPINE

A35
OPHTHALMIC DRUG DELIVERY SYSTEMS
Fraunfelder, F.T. & Hanna, C.
Surv. Ophthalmol., 18, 1974, p. 292
PILOCARPINE

A36
EXPERIENCE WITH THE USE OF ACTIVE AND INERT INTRAUTERINE DEVICES
AND THEIR ACTION ON THE UTERUS, ENDOMETRIUM, AND VAGINAL
CYTOLOGY
Ruiz-Velasco, V. & Illanes-Amurrio, S.
J. Steroid Biochem., 5, 1974, p. 377
PROGESTERONE

A37
DIFFUSIONAL SYSTEMS FOR CONTROLLED RELEASE OF DRUGS TO THE EYE
Shell, J.W. & Baker, R.W.
Ann. Ophthalmol., 6, 1974, p. 1037
PILOCARPINE

A38
EFFECT OF INTRAUTERINE PROGESTERONE ON THE HYPOTHALAMIC-PITUITARY
OVARIAN AXIS IN BABOONS
Tillson, S.A. et al.
J. Steroid Biochem., 5, 1974, p. 376

A39
COMPLICATION OF THE USE OF OCUSERTS
Abrahamson, I.A.
Arch. Ophthalmol., 93, 1975, p. 317
PILOCARPINE

A40
CLINICAL STUDY OF A PROGESTERONE-RELEASING INTRAUTERINE
CONTRACEPTIVE DEVICE
Brenner, P.F. et al.
Am. J. Obstet. Gynecol., 121, 1975, p. 704

A41
UTERINE THERAPEUTIC SYSTEM: A NEW APPROACH TO FEMALE
CONTRACEPTION
Connell, E.B.
Contemp. Ob./Gyn., 6, 1975, p. 49
PROGESTERONE

A42
DURATION OF ACTION OF PILOCARPINE OCUSERT ON INTRAOCULAR
PRESSURE IN MAN
Drance, S.M. et al.
Can. J. Ophthalmol., 10, 1975, p. 450

A43
LONG-ACTING OCUSERT-PILOCARPINE SYSTEM IN THE MANAGEMENT OF
GLAUCOMA
Lee, P.F. et al.
Invest. Ophthalmol., 14, 1975, p. 43

A44
ENDOMETRIAL MORPHOLOGY IN WOMEN EXPOSED TO UTERINE SYSTEMS
RELEASING PROGESTERONE
Martinez-Manautou, J. et al.
Am. J. Obstet. Gynecol., 121, 1975, p. 175

A45
OCUSERT SYSTEM IN THE MANAGEMENT OF GLAUCOMA
Novak, S. & Stewart, R.H.
Tex. Med., 71, 1975, p. 63
PILOCARPINE

A46
LOCAL EFFECTS OF TOPICALLY APPLIED STEROIDS
Place, V. & Benson, H.
J. Steroid Biochem., 6, 1975, p. 717
PILOCARPINE; PROGESTERONE

A47
PILOCARPINE OCUSERTS. LONG-TERM CLINICAL TRAILS AND SELECTED
PHARMACODYNAMICS
Quigley, H.A. et al.
Arch. Ophthalmol., 23, 1975, p. 771

A48
OCULAR MICROTHERAPY. MEMBRANE-CONTROLLED DRUG DELIVERY
Richardson, K.T.
Arch. Ophthalmol., 93, 1975, p. 74
PILOCARPINE

A49
COMPARATIVE DISTRIBUTION OF PILOCARPINE IN OCULAR TISSUES OF THE
RABBIT DURING ADMINISTRATION BY EYEDROP OR BY MEMBRANE-
CONTROLLED DELIVERY SYSTEMS
Sendelbeck, L. et al.
Am. J. Ophthalmol., 80, 1975, p. 274

A50
EFFECT OF INTRAUTERINE PROGESTERONE ON THE HYPOTHALAMICHYPOPHYSEAL-OVARIAN AXIS IN HUMANS
Tillson, S.A. et al.
Contraception, 11, 1975, p. 179

A51
INDUCTION OF HUMAN ENDOMETRIAL ESTRADIOL DEHYDROGENASE BY PROGESTINS
Tseng, L. & Gurpide, E.
Endocrinology, 97, 1975, p. 825
PROGESTERONE

A52
NOVEL METHODS OF OCULAR DRUG DELIVERY
Urquhart, J.
Proc. 6th Int'l. Cong. Pharmacol., 5, 1975, p. 63
Tuomisto, J. & Paasonen, M.K. eds.
PILOCARPINE

A53
THERAPEUTIC IMPLICATIONS OF CONTROLLED DRUG DELIVERY
Zaffaroni, A.
Proc. 6th Int'l. Cong. Pharmacol., 5, 1975, p. 53
PILOCARPINE; PROGESTERONE

A54
PILOCARPINE OCUSERTS: A CLINICAL TRIAL OF EFFICACY, WEAR-TOLERANCE AND PATIENT BENEFIT
Goldberg, I. & Hollows, F.C.
Aust. J. Ophthalmol., 4, 1976, p. 168

A55
PROGESTERONE-RELEASING INTRAUTERINE CONTRACEPTIVE DEVICE
Gyozo, G.
Am. J. Obstet. Gynecol., 124, 1976, p. 214

A56
MODES OF ACTION OF MEDICATED INTRAUTERINE DEVICES
Hagenfeldt, K.
J. Reprod. Fert. Suppl., 25, 1976, p. 117
PROGESTERONE

A57
BIOLOGICAL AND CLINICAL CONSIDERATIONS ON INTRAUTERINE USE OF
PROGESTERONE-T FOR FERTILITY CONTROL
Leone, U. et al.
Acta Eur. Fertil., 7, 1976, p. 145

A58
EFFECT OF T-SHAPED PROGESTERONE-RELEASING SYSTEM ON CONTRACEPTIVE
EFFICACY AND MENSTRUAL CYCLE REGULARITY IN JAPANESE MONKEYS
Oshima, K. et al.
Fertil. Steril., 27, 1976, p. 582

A59
COMPLICATION WITH THE USE OF OCUSERT
Pearson, D.C.
Arch. Ophthalmol., 94, 1976, p. 168
PILOCARPINE

A60
STEROID DELIVERY SYSTEMS FOR CONTRACEPTION
Pharriss, B.B. et al.
J. Reprod. Med., 17, 1976, p. 91
PROGESTERONE

A61
OCUSERT PILOCARPINE SYSTEM: ADVANTAGES AND DISADVANTAGES
Pollack, I.P.
S. Med. J., 69, 1976, p. 1296

A62
INTERRELATIONSHIPS BETWEEN PITUITARY GONADOTROPHINS AND OVARIAN
STEROIDS IN BABOONS DURING CONTINUOUS INTRAUTERINE PROGESTERONE
TREATMENT
Tillson, S.A. et al.
Biol. Reprod., 15, 1976, p. 291

A63
PROGESTASERT IUD
Bounds, W.
Fertil. Contracep., 1, 1977, p. 37
PROGESTERONE

A64
HORMONAL EVALUATION OF THE INTRAUTERINE PROGESTERONE
CONTRACEPTIVE SYSTEM
Bryant-Greenwood, G.D. et al.
J. Clin. Endocrinol. Metab., 44, 1977, p. 721

A65
BIOCHEMICAL AND MORPHOLOGICAL CHANGES IN THE HUMAN
ENDOMETRIUM INDUCED BY THE PROGESTASERT DEVICE
Hagenfeldt, K. et al.
Contraception, 16, 1977, p. 183
PROGESTERONE

A66
MANAGEMENT OF CHRONIC GLAUCOMA WITH PILOCARPINE OCUSERTS
Hillman, J.S. et al.
Trans. Ophthal. Soc. U.K., 97, 1977, p. 206

A67
EXPERIENCE WITH PILOCARPINE OCUSERTS
Hitchings, R.A. & Smith, R.J.H.
Trans. Ophthal. Soc. U.K., 97, 1977, p. 202

A68
THERAPEUTIC EFFECTS OF PROGESTASERT
Kent, S.
Contemp. Ob./Gyn., 10, 1977, p. 33
PROGESTERONE

A69
INTRAUTERINE DEVICES CONTAINING PROGESTERONE
Murad, F.
Drug Therapy, 7, 1977, p. 119

A70
COMPARATIVE STUDY OF THE PROGESTASERT T AND COPPER 7 IUD
Pizarro, E. et al.
Contraception, 16, 1977, p. 313
PROGESTERONE

A71
HORMONE-RELEASING IUDs — A PRELIMINARY ASSESSMENT
Snowden, R.
Fertil. Contracep., 1, 1977, p. 3
PROGESTERONE

A72
EFFECTS OF A PROGESTERONE-T INTRAUTERINE DEVICE ON CARBOHYDRATE
AND LIPID METABOLISM IN 'NORMAL' WOMEN: A THREE MONTH STUDY
Spellacy, W.N. et al.
Contraception, 15, 1977, p. 65

A73
EFFECTS OF THE PROGESTASERT ON THE MENSTRUAL PATTERN, OVARIAN
STEROIDS AND ENDOMETRIUM
Wan, L.S. et al.
Contraception, 16, 1977, p. 417
PROGESTERONE

A74
COMPARATIVE ULTRASONOGRAPHY STUDY OF THE EFFECT OF PILOCARPINE
2% AND OCUSERT P20 ON THE EYE COMPONENTS
Francois, J.
Amer. J. Ophthalmol., 86, 1978, p. 233

A75
INTRAUTERINE DEVICES AND MENSTRUAL BLOOD LOSS: A COMPARATIVE
STUDY OF EIGHT DEVICES DURING THE FIRST SIX MONTHS OF USE
Gallegos, A.J. et al.
Contraception, 17, 1978, p. 153
PROGESTERONE

A76
NO INCREASE IN THE FIBRINOLYTIC ACTIVITY OF THE HUMAN ENDOMETRIUM
BY PROGESTERONE RELEASING IUD (PROGESTASERT)
Liedholm, P. et al.
J. Ster. Biochem., 9, 1978, p. 850

A77
OCUSERT IN OPEN-ANGLE GLAUCOMA
Marner, O.
Acta Ophthalmol., 56, 1978, p. 483
PILOCARPINE

A78
STUDIES ON THE INFLUENCE OF OCUSERT P40 OR PILOCARPINE EYE DROPS
(2%), RESPECTIVELY ON THE RABBIT LENS
Ohrloff, C. et al.
Ophthalmic Res., 10, 1978, p. 324

A79
ENDOMETRIAL PROSTAGLANDIN F CONTENT IN WOMEN WEARING NON-
MEDICATED OR PROGESTIN-RELEASING INTRAUTERINE DEVICES
Scommegna, A. et al.
Fertil. Steril., 29, 1978, p. 500
PROGESTERONE

A80
STUDIES OF CARBOHYDRATE AND LIPID METABOLISM IN WOMEN USING THE
PROGESTERONE-T, INTRAUTERINE DEVICE FOR SIX MONTHS: BLOOD GLUCOSE,
INSULIN, CHOLESTEROL, AND TRIGLYCERIDE LEVELS
Spellacy, W.N. et al.
Fertil. Steril., 29, 1978, p. 505

A81
INTRODUCTION OF THE OCUSERT OCULAR DELIVERY SYSTEM TO AN
OPHTHALMIC PRACTICE
Stewart, R.H. & Novak, S.
Ann. Ophthalmol., 10, 1978, p. 325
PILOCARPINE

A82
OCUSERT IN ACUTE GLAUCOMA
Vase, I. & Marner, O.
Acta Ophthalmol., 56, 1978, p. 483
PILOCARPINE

A83
PILOCARPINE MEDICATION IN OPEN-ANGLE GLAUCOMA
Brinchmann-Hansen, O. & Anmarkrud, N.
Acta Ophthalmol., 57, 1979, p. 55

A84
EXPERIENCE WITH TWO DIFFERENT MEDICATED INTRAUTERINE DEVICES: A
COMPARATIVE STUDY OF THE PROGESTASERT AND NOVA-T
Fylling, P. & Fagerhol, M.
Fertil. Steril., 31, 1979, p. 138
PROGESTERONE

A85
COMPARISON OF THE PUPILLARY, REFRACTIVE, AND HYPOTENSIVE EFFECTS OF
OCUSERT-40 AND PILOCARPINE EYEDROPS IN THE TREATMENT OF CHRONIC
SIMPLE GLAUCOMA
Smith, S.E. et al.
Br. J. Ophthalmol., 63, 1979, p. 228

A86
CARBOHYDRATE AND LIPID STUDIES IN WOMEN USING THE PROGESTERONE
INTRAUTERINE DEVICE FOR ONE YEAR
Spellacy, W.N.
Fertil. Steril., 31, 1979, p. 381

A87
INTRAUTERINE PROGESTERONE CONTRACEPTIVE SYSTEM AS AN ALTERNATIVE
IN CASES WHERE CONVENTIONAL IUD's ARE UNSUITABLE
Ylostalo, P. et al.
Acta Obstet. Gynecol. Scand., 58, 1979, p. 279

A88
CORPUS LUTEUM FUNCTION IN LACTATING WOMEN USING PROGESTASERT
SYSTEM
Abdalla, M.I. et al.
Contracept. Deliv. Syst., 1, 1980, p. 259
PROGESTERONE

A89
MULTICENTRE STUDY ON THE EFFECT AND TOLERANCE OF OCUSERT-P40
Akerblom, T. et al.
Acta Ophthalmol., 58, 1980, p. 617
PILOCARPINE

A90
EFFECT OF INTRAUTERINE DELIVERY OF PROGESTERONE ON THE SECRETION
OF UTEROGLOBULIN
Allen, J.L. & Fang, S.M.
Fed. Proc., 39, 1980, p. 983

A91
PROGESTASERT SYSTEM: PITUITARY ADRENAL FUNCTION
Askalini, H. et al.
Contracept. Deliv. Syst., 1, 1980, p. 261
PROGESTERONE

A92
PROGESTASERT IN THE FIRST YEAR POST-PARTUM
Badraoui, M.H.H. et al.
Contracept. Deliv. Syst., 1, 1980, p. 192
PROGESTERONE

A93
SERUM PROLACTIN LEVEL IN LACTATING WOMEN USING PROGESTASERT
SYSTEM
Badraoui, M.H.H. et al.
Contracept. Deliv. Syst., 1, 1980, p. 260
PROGESTERONE

A94
PROGESTERONE INTRAUTERINE DEVICE, ITS USES AND EFFICIENCY
Cittadini, E. et al.
Acta Europaea Fertil., 11, 1980, p. 199

A95
THYROID PROFILE IN LACTATING WOMEN USING THE PROGESTASERT
Ibrahim, I.I. et al.
Contracept. Deliv. Syst., 1, 1980, p. 260
PROGESTERONE

A96
STUDY OF THE PROGESTASERT INTRAUTERINE DEVICE IN GENERAL PRACTICE
London, N.B.
Br. J. Fam. Plan., 6, 1980, p. 40
PROGESTERONE

A97
OCUSERT PILOCARPINE THERAPEUTIC SYSTEM IN GLAUCOMA MANAGEMENT.
(A REVIEW)
Shihab, Z. & Lee, P.F.
Perspect. Ophthalmol., 4, 1980, p. 79

A98
PRACTICAL PROBLEMS IN THE USE OF OCUSERT-PILOCARPINE DELIVERY SYSTEM
Sihvola, P. & Puustjarvi, T.
Acta Ophthalmol., 58, 1980, p. 933

GLUTAMIC ACID POLYMER

A99
BIODEGRADABLE, IMPLANTABLE SUSTAINED RELEASE SYSTEMS BASED ON
GLUTAMIC ACID COPOLYMERS
Sidman, K.R. et al.
J. Memb. Sci., 7, 1980, p. 277
PROGESTERONE; NORETHISTERONE; NORGESTREL; NALTREXONE; PLATINUM;
PRIMAQUINE DIPHOSPHATE

LACTIC ACID POLYMER

A100
BIODEGRADABLE MICROSPHERE CONTRACEPTIVE SYSTEM
Beck, L.R. & Cowsar, D.
Acta Europ. Fertilitatis, 11, 1980, p. 139
NORETHISTERONE; PROGESTERONE; ESTRADIOL

PHENYLENE OXIDE POLYMER

A101
MICROCAPSULES FOR CONTROLLED RELEASE VIA DONNAN DIALYSIS
Cabasso, I.
J. Memb. Sci., 7, 1980, p. 305
SODIUM NITRATE; POTASSIUM BICARBONATE; SODIUM CHLORIDE; SODIUM
PHOSPHATE

SILICONE RUBBER

A102
BIOAVAILABILITY OF NORETHINDRONE IN RABBITS AFTER ADMINISTRATION OF NORETHINDRONE ACETATE IN SINGLE, DOUBLE AND QUADRUPLE DOSES RELEASED THROUGH SUBCUTANEOUS SILASTIC IMPLANTS
Sarkar, N.N. et al.
Endocrinol. Jpn, 27, 1980, p. 653

A103
AN OESTRONE-RELEASING VAGINAL RING IN THE TREATMENT OF CLIMACTERIC WOMEN
Sipinen, S. et al.
Maturitas, 2, 1980, p. 291

A104
EFFECT OF NORETHISTERONE ENANTHATE SUBDERMAL SILASTIC IMPLANTS ON THE OVARY AND UTERUS OF ADULT RABBITS
Srivastava, U.K. & Mendiratta, R.
Indian, J. Exp. Biol., 18, 1980, p. 1221

INDEX OF POLYMERS AND COPOLYMERS

VINYL PYRROLIDONE POLYMERS AND COPOLYMERS
Vinyl Pyrrolidone Polymer (PVP, Povidone, Percoll)   1138
Vinyl Pyrrolidone-Maleic Anhydride Copolymer   1280
Vinyl Pyrrolidone-Methacrylic Acid Copolymer   1280

VINYL SULPHATE POLYMER   1281

VINYL SULPHONIC ACID POLYMER   1284

XENYL PHOSPHATE POLYMER   1285

INDEX OF ACTIVE AGENTS

Mitomycin   179, 180,
  399, 400, 632, 633, 635,
  935, 995, 1130
Monochloroethylamine   1280,
  1281
Morphine   742, 857, 884,
  910, 1156
Morpholine   268

N
Naloxone   337, 499, 500,
  505, 508, 857
Naltrexone   337, 499,
  500, 502, 505, 506, 507,
  508, 509, 518, 521, A99
Beta-naphthol   356, 688, 1162
Neomycin   150, 161, 422,
  667, 1238
Neostigmine   280
Nicotine   779, 780
Nicotinic acid   1124, 1248
Niridazole   908
Nitrofurantoin (Furadantin)   988
Nitroglycerin   393, 1241,
  1259
Nitrosourea   755, 771,
  810, 865, 883
Nitrous oxide   700
Norephedrine   330, 985
Norethandrolone   863
Norethindrone   61, 243, 273,
  275, 295, 320, 341, 348,
  440, 478, 515, 701, 723,
  794, 798, 800, 806, 807,
  830, 831, 835, 836, 838,
  843, 850, 854, 858, 859,
  863, 867, 872, 886, 889,
  895, 920, A102
Norethisterone   485, 503,
  513, 523, 880, 881, 893,
  A99, A100, A103, A104
Norgestomet   552, 814
Norgestrel   61, 273, 341,
  348, 440, 485, 513, 515,
  517, 703, 723, 724, 732,

735, 764, 798, 800, 806,
807, 822, 825, 827, 828,
831, 835, 841, 861, 862,
863, 864, 868, 872, 876,
877, 878, 880, 894, 895,
898, 906, 907, 911, 915,
A99
Norgestrienone   776, 863, 864
Norprogesterone   61, 243,
  275, 295, 320, 341, 440,
  478, 701, 703, 714, 721,
  723, 806, 920
Nortestosterone   732

O
Oestrogen   715
Oestrone (Estrone)   680, 714,
  816, 846, 855, 914, A103
Organotin halides   411
Orphenadrine   1051, 1129
Oxalic acid   1153
4-Oxycumarine   1248
Oxprenolol   1011

P
Papain   129, 159, 259,
  996, 1026
Papaverine   616, 978
Paracetemol   216
Parahydroxybenzoic acid esters
  (Parabens)   247, 248, 356,
  357, 358, 367, 688, 787,
  1159, 1162, 1177
Penicillin   1154, 1158
Pentobarbital   181, 182, 245,
  257, 616, 978
Pentobarbitone   1155
Pepsin   226, 260
Pethidine   1156
Phenazone (Antipyrine)   219
Phenethylamine   652, 983, 1221
Phenobarbital   360, 527,
  676, 940, 987, 1017, 1159

AUTHOR INDEX

Numbers in italics and underlined refer to editorship

Graffner, C.   325, 326, 334,
  979, 980, 989, 1215
Graham, N.B.   104, 398, 954
Graham, W.D.   1155, 1156,
  1157
Greenberg, L.   1180
Greene, W.A.   855
Gregoriadis, G.   _110_, _486_
Gresser, J.D.   523
Grimova, J.   665
Gruber, U.F.   1208
Guess, W.L.   1034
Gupta, G.N.   802
Gurpide, E.   A51
Guttman, D.E.   355, 356, 688,
  1160, 1162
Gyozo, G.   A55

H
Hass, J.S.   1001
Habal, M.B.   630, 631, 901,
  902, 959, 960
Hadgraft, J.   111
Hafez, E.S.E.   _38_, _41_, _A16_, _A17_
Hagenfeldt, K.   437, A56, A65
Haldon, R.A.   560
Haleblian, J.   743
Haler, D.   952
Hall, C.E.   1006
Hall, O.   1006
Halpern, B.D.   537,   546, 576,
  621, 1242
Hamacher, H.   933
Handa, T.   224
Hanna, C.   A35
Hanson, S.R.   156
Hart, P.D.   371
Heilmann, K.   A7, A8
Heller, J.   120, 526,
  998, 1128
Hendry, R.M.   261, 934
Henzl, M.R.   763
Herrman, J.E.   261, 934
Hess, H.J.   _A21_

Heyd, A.   412, 419, 420,
  422, 667
Hickman, R.K.   1272
Hicks, G.P.   131
Hiestand, E.N.   6
Higashide, F.   925
Higgins, H.P.   1195
Higuchi, T.   8, 72, 355, 356,
  688, 1158, 1159, 1160, 1162
Higuchi, W.I.   6, 7, 15, 731
Hillier, S.G.   836
Hillman, J.S.   1228, A66
Hindberg, I.   567, 1229
Hirano, T.   1100
Hirsch, M.S.   1063, 1074, 1077
Hitchings, R.A.   A67
Hixson, H.F.   _73_
Hochster, R.M.   _5_
Hofmann, V.   948
Holland, J.F.   303, 304
Hollows, F.C.   A54
Holt, P.F.   1134, 1135, 1136
Hopfenberg, H.B.   74
Horejsi, V.   A25
Horne, H.W.   720
Hosaka, S.   168, 627, 636, 1261
Hosono, T.   1273
Hrabak, F.   665
Hsia, H.T.   672
Hsiao, C.H.   1247
Huang, S.J.   300
Huberman, H.S.   837
Huberman, S.   1034
Hueper, W.C.   695, 961, 1000,
  1165, 1167, 1181
Hummel, J.P.   1286
Hung, G.W.C.   965

I
Ibrahim, I.I.   A95
I.C.C.R.P.C.   856
Ichimura, K.   1052, 1274
Ida, T.   614, 947
Iliev, I.   53, 1262, 1280, 1281

Yoshida, M.
171, 172, 175, 178, 185,
539, 540, 541, 547, 549,
556, 590, 598, 599, 600,
607, 608, 610, 626, 641,
642, 643, 942
Yunker, M.H.   280
Yurdakul, S.   581

Z
Zaborsky, O.   42
Zador, G.   451

Zaffaroni, A.
35,   84,   94,   127, 128,
215, 442, 1022, 1260, A11,
A12, A24, A53
Zaharko, D.S.   115
Zander, A.R.   1115, 1116
Zatz, J.L.   1123, 1217, 1222
Zbuzkova, V.   735, 736
Zeller, H.   1265
Zentner, G.M.
550, 591, 592, 601, 602,
611, 612, 639, 869, 956,
970
Zilkha, A.   375
Zimecki, M.   1279